# PHOTOELECTRIC PHOTOMETRY OF VARIABLE STARS

Second, Revised and Expanded Edition

## A Practical Guide for the Smaller Observatory

Douglas S. Hall, Russell M. Genet

## Willmann–Bell, Inc.

P.O. Box 35025
Richmond, Virginia 23235
United States of America

Publishers and Booksellers

Serving Astronomers Worldwide
Since 1973

First published 1982
Second, Expanded and Enlarged Edition 1988


Printed in the United States of America




**Library of Congress Cataloging-in-Publication Data.**
Hall, Douglas S., 1940–
    Photoelectric photometry of variable stars: a practical guide for
the smaller observatory    Douglas S. Hall, Russell M. Genet wth
contributions by I.A.P.P.P. members – 2nd, rev., and expanded ed.
        p.    cm.
    Bibliography: p.
    Includes index.
    ISBN 0-943396-19-0 :
    1. Stars, Variable–Observations. Photometry, Astronomical-
-Observations. I. Genet, Russell. II. International Amateur
-Professional Photoelectric Photometry (organization) III. Title.
QB836.H35  1989
523.8'44- -dc 19                                        88-10822
                                                            CIP

# PREFACE

Our objective in writing this book was to provide in one convenient source most of the information smaller observatories need to equip themselves for and conduct photoelectric observations of variable stars. The entire field of astronomical photoelectric photometry was too extensive to treat in a single book, so we chose to treat only the photometry of variable stars. We have included material we feel is of special interest to smaller observatories, especially those smaller observatories just getting started in photometry. Thus, our emphasis is on the practical considerations and details of observatories, telescopes, photoelectric equipment, observing techniques, data reduction, and planning of observing programs. The 1963 book *Photoelectric Astronomy for Amateurs*, by F. Bradshaw Wood, has been the inspiration for and guide to photoelectric photometry at smaller observatories for almost twenty years. In recognition of Dr. Wood's many contributions to astronomical photometry and his fostering of small-observatory photometry, we asked him to write the Foreword to this book, and are grateful that he did so. Astronomical photoelectric photometry is almost entirely a development of this century, and it has been greatly influenced by technological advancements. We consider an understanding of the history of photoelectric photometry to be a necessary ingredient of every photometrist's education. Gerald E. Kron played a key role in the history of photometry, and personally knew the other principle developers of astronomical photoelectric photometry. We deeply appreciate the definitive chapter on the history of photoelectric photometry he has provided. We wish to acknowledge our deep gratitude to the many amateur and professional astronomers who have so generously contributed sections to this book. Our very special thanks are also due to Arne Henden, Ronald Kaitchuck, Kevin Krisciunas, Howard Louth, and T. M. Young who took the time to review and comment in detail on the prepublication copy of this book, and to the many others who provided helpful suggestions. Our thanks also to G. Bower, C. Boardner, O. Brettman, R. Fried, W. Hanson, J. Hopkins, J. Johnson, R. Lines, H. Louth, T. Renner, H. Stelzer, F. Tretta, and M. Zeilik for providing the photographs in the chapter on photoelectric observatories. Finally, this book is dedicated to Mimi and Joyce, constant companion stars of the first magnitude.

Douglas S. Hall
Dyer Observatory
Vanderbilt University
Nashville, TN 37235
Phone (615) 373-4897

Russell M. Genet
Fairborn Observatory
1750 S. Price #127
Tempe, AZ 85281
Phone (602) 968-3899

# FOREWORD

No other branch of science has profited as much from the contributions of diligent and intelligent amateurs than has the study of astronomy. The Herschel family comes immediately to mind, and John Goodricke as well, but there have been many others.

The American Meteor Society, it is true, had for many years a professional, Charles P. Olivier, as President but other than that was composed almost exclusively of amateurs. Many discoveries of comets were made by amateur astronomers; others have served as volunteer aides on solar eclipse expeditions. But in no branch of astronomy have they served as well as in the study of variable stars. The BBSAG (Switzerland), the AAVSO (United States), and many similar associations in other nations have made and published observations of variable stars of various types. These observations have been almost entirely photometric since few amateurs could afford a spectrograph and a telescope of the size needed to use one effectively.

Until fairly recently in the long history of astronomy, these observations were made by visual estimates. This in no way limited their value since most professional astronomers were using either visual or photographic estimates and, when made by an experienced observer, the visual were at least as good as the photographic. A very few visual polarizing or wedge photometers were in use and these indeed were free from some of the errors that affected estimates; very rarely there appeared photographic work from measures (as distinct from estimates); photoelectric photometers were temperamental instruments which frequently could be maintained only by the astronomer who had himself designed and built them.

The situation changed dramatically with the introduction of the multiplier photocell which was developed during World War II. Another advantage—one rarely mentioned—was the increased reliability of electrical components available. The use of the multiplier photocell made it possible to work with a 12-inch telescope to limits previously available only with one of 60 inches or larger. The limiting factor now was the sky background.

It was not long before a few amateurs began building and using photoelectric photometers and the number increased with the years. Nearly twenty years ago, the number was such that it seemed in order to publish a small book describing both the construction of a simple photometer (with the electronic equipment then available) and its use. Both the increase in the number of amateur astronomers and the sophistication of the equipment now avaiable have suggested that another, somewhat more extensive, volume is now needed. This is also suggested by the growing collaboration between amateurs and professionals as indicated, for example, by the formation of I.A.P.P.P.

One of the most important decisions any astronomer can make is the selection of a worthwhile observing program. This is discussed in more detail later in the text and only a few remarks will be made here. In general, unless the astronomer is willing to take a good deal of care in calibrating and checking a color system, it would be wise to choose a program not very sensitive to color. One example is the study of the times of minima of eclipsing binaries. Indeed someone has said that a sure way to immortality is to determine and to publish such times of minima, for they will be used in all future studies.

It is true that until the 1930's, and even later, mention was made of the Tikhoff-Nordman effect. This was the occurrence of minima at slightly different times when observed in different colors, supposedly caused by light of different wavelengths traveling at slightly different velocities. However, the classical study of Algol by John Hall, the compilation by Rosa Szafraniec of a large sample of photoelectrically observed minima, and a host of later evidence have showed rather clearly that this effect does not exist, and the observed times of minima will be insensitive to color. Further, significant results can often be obtained from one night's observing.

There are, of course, many other useful programs and, as previously stated many of these are discussed later in this book. The main point is that the amateur astronomer with a photoelectric photometer now can carry out work which will be of scientific significance and of permanent value.

Frank Bradshaw Wood
Department of Astronomy
University of Florida

# Table of Contents

# Chapter 1

# INTRODUCTION

## 1.1    Photoelectric Photometry of Variable Stars

While the light of most stars is quite steady, there are other stars whose light varies over time. There are two broad classes: those that are intrinsically variable, i.e., they vary in brightness due to various mechanisms within a single star; and extrinsically variable stars, where the cause of variability is external to the star itself, such as eclipsing binaries.

Variable stars are of special interest to astronomers because the variability provides much information on stars that would not otherwise be obtainable. The period-luminosity relationship of Cepheid variables, for instance, has made available a valuable means of determining distances. Eclipsing binaries have provided much valuable information on stellar masses and diameters. Long period variables have provided information that has greatly increased our understanding of stellar evolution. These are just a few of the many contributions that have been made and will continue to be made by measuring the light of variable stars.

There are a number of ways that the changing intensity of starlight over time can be measured. These include visual measurements (with or without special equipment aids), photographic measurements, and photoelectric measurements. Visual estimates of the intensity of light from a variable star can be made without any equipment other than a telescope by comparing the light of the variable star with other nearby non-variable stars. Special finder charts and instructions for this approach to observing variable stars can be obtained from the American Association of Variable Star Observers (AAVSO) and other organizations. The address of the AAVSO is given in Appendix B.

The primary advantages of visual estimates are that they require no special equipment, the estimates can be made quite rapidly, and the requirements for specialized training and extensive data analysis are minimal. Unaided visual estimates are necessarily limited in accuracy to about 0.2 of a magnitude, and to the visible portion of the spectrum, as one cannot see in the ultraviolet or infrared. In spite of these limitations and the somewhat variable color response between the eyes of different observers, visual estimates continue to be of considerable use. This is particularly true for longer period variable stars with sizable changes in light intensity over time, and for cataclysmic variables. For these types of stars, the sheer number of visual observers and the efficiency with which they can make observations allows coverage of many hundreds of variable stars that would oth-

1

erwise go unobserved due to the limited number of observatories equipped for the other, more accurate forms of measuring the light of variable stars.

In the 19th century, a number of aids were developed to improve the accuracy of visual observations. These included many methods for attenuating the light of a variable star in a calibrated manner so that it could be compared more directly with a reference star. These approaches have generally fallen into disuse as photographic and photoelectric measurements were found to be superior.

Photographic photometry is based on the relationship between the intensity of the light from a star, and the size and/or photographic density of the stellar image on the developed plate. This relationship, unfortunately, is not a simple one and varies from one batch of plates to the next. It is also dependent to some extent on the conditions under which the plates were exposed. For these reasons, photographic photometry is usually assisted by photoelectric photometry. Photoelectric photometry is used to measure separately the intensity of a number of stars over a range of magnitudes in the field covered by the plate, thereby providing a quantitative calibration. Sometimes, a photoelectric photometer (usually referred to as a microdensitometer) is used to measure the intensity of a controlled light source that passes through each star image on the developed plate. Photographic photometry can thus be a somewhat involved process requiring calibration, measurements, and analysis, as well as assistance for photoelectric photometry. However, if an accuracy of 0.1 magnitude is sufficient, then photographic plates can be inspected visually, comparing the variable star with a sequence of nearby non-variable stars, much as is done in visual photometry.

Photographic photometry has two considerable advantages. First, in a single exposure it can cover many stars at once. Second, due to the accumulation of light over time, it can reach stars much fainter than those that can be visually located in the telescope. For these reasons, photographic photometry has proven to be particularly valuable in the study of variable stars in faint open and globular clusters and in nearby galaxies. Photographic photometry nicely compliments visual and photoelectric photometry. The Maria Mitchel Observatory, under the directorship of Emilia P. Belserene, is one of the few observatories continuing this valuable form of variable star research. It is a slightly sad comment on amateur astronomy that, of the thousands of cameras and tens of thousands of rolls of film used to photograph Messier objects for the hundred-thousandth time, not a single amateur camera (we know of) is trained regularly on clusters to track the variations of the RR Lyrae-type and other scientifically interesting variables.

Photoelectric photometry is simply the electrical measurement of the light intensity of individual stars or other objects. A photoelectric photometer is essentially a very sensitive exposure meter similar in principle to those used in ordinary photography. There are three principle advantages to photoelectric photometry of variable stars. The first is its accuracy, which is considerably greater than that currently obtainable by any other approach. The second is that electro-optical sensors, unlike the photographic emulsion and the human eye, can be made to respond linearly, i.e., in direct proportion, to the rate at which the luminous energy falls upon them. The third is the wide wavelength range, from the ultraviolet out into the infrared, that can be covered.

Like photographic photometry, photoelectric photometry suffers from the requirements of specialized equipment, calibration, detailed analysis, and specialized knowledge. Unlike photographic photometry, but similar to visual photometry, photoelectric photometry is generally limited to measuring one star at a time, and in general the star must be bright enough to be seen. In spite of these not inconsiderable disadvantages, photoelectric photometry has been the primary tool of professional astronomers for measuring the intensity of the light of variable stars for over 30 years. Most of the major observatories have telescopes (usually their smaller ones) that are almost totally dedicated to photoelectric photometry. Many of the smaller professional observatories devote most of their observational time to photoelectric photometry.

## 1.2 Photoelectric Photometry and the Small Observatory

In this age of interplanetary probes, orbiting telescopes, and gigantic radio and optical telescopes, it seems pertinent to ask if any serious research can be done using small, ground-based optical telescopes. A sizable number of colleges and advanced amateurs have smaller telescopes of 0.4- meters (16 inches) aperture or less and many of their proprietors would like to use them for serious research.

One way of defining "serious research" is the operational definition of "that which is published in leading journals." To gauge what types of serious observations are being made with small ground-based optical telescopes, the *Astronomical Journal* was surveyed by one of us (Genet) from 1975–1979. All papers that were based on observations from at least one telescope of an aperture of 0.4 meter (16 inches) or less were reviewed. The results of the survey are shown in Table 1.1.

While this survey was certainly not exhaustive and may not be truly representative of papers published in all leading astronomical journals, the data shown in Table 1.1 were subjected to some simple analysis. Of the 30 papers, eight were based on observations from at least one telescope of 0.3 meters (12 inches) or less. While small telescopes at large institutions, such as Kitt Peak National Observatory, were used to make the observations for 21 of the papers, nine of the observations were made at least in part by amateurs or at small colleges. The observational technique in 28 of the 30 papers was photoelectric photometry, while one was photographic and one was spectroscopic. Of the 28 papers based on photoelectric observations, 18 used the UBV system, six used the uvby$\beta$ system, and four used other photometric systems.

Of the 28 papers based on photoelectric observations, nine were of eclipsing binaries, five were of stellar color (such as cluster magnitude-color arrays), five were of asteroids (or comets), and three each were of intrinsically variable stars, planetary or asteroid occultations, and miscellaneous.

Based on this admittedly limited survey, it appeared that serious astronomical research is indeed being done with optical ground-based telescopes of 0.2- to 0.4- meter aperture (8 to 16 inches). These smaller telescopes were being utilized by advanced amateurs, small colleges, and large institutions almost entirely for photoelectric photometry. While eclipsing binaries were most frequently observed, considerable research in the areas of stellar color and asteroids was also reported, with slightly less reported in the areas of intrinsic variable stars and planetary occultations.

| Table 1.1 |
|:---:|
| Summary of 30 papers in the *Astronomical Journal* |
| based on observations made with small telescopes |

| (pe = photoelectric, pg = photographic, sp = spectrographic) |
|:---:|

| Vol. | Page | Aperture | Observatory | Method | Observations |
|:---:|:---:|:---:|:---|:---:|:---|
| 80 | 56 | 16″ | Agassiz | pe | planetary occ. |
| 80 | 128 | 16″ | K.P.N.O. | pe | stellar color |
| 80 | 140 | 16″ | C.T.I.O. | pe | eclipsing binary |
| 80 | 246 | 7″ | Steward | pg | comets & asteroids |
| 80 | 637 | 16″ | C.T.I.O. | sp | stellar class. |
| 80 | 698 | 16″ | K.P.N.O. | pe | intrinsic variable |
| 80 | 852 | 12″ | U. of MD | pe | comet |
| 81 | 57 | 15″ | Villanova | pe | eclipsing binary |
| 81 | 67 | 16″ | K.P.N.O. | pe | asteroid |
| 81 | 74 | 16″ | C.T.I.O. | pe | asteroid |
| 81 | 107 | 16″ | K.P.N.O. | pe | visual binaries |
| 81 | 182 | 16″ | K.P.N.O. | pe | standard stars |
| 81 | 228 | 16″ | K.P.N.O. | pe | stellar color |
| 81 | 632 | 16″ | C.T.I.O. | pe | stellar color |
| 81 | 661 | 15″ | Flower & Cook | pe | intrinsic variable |
| 81 | 665 | 16″ | N. GA Col. | pe | eclipsing binary |
| 81 | 778 | 16″ | K.P.N.O. | pe | asteroid |
| 81 | 885 | 16″ | K.P.N.O. | pe | eclipsing binary |
| 83 | 176 | 8″ | Landis | pe | eclipsing binary |
| 83 | 201 | 16″ | K.P.N.O. | pe | asteroid |
| 83 | 278 | 16″ | C.T.I.O. | pe | stellar color |
| 83 | 438 | 12″ | Beverly Begg | pe | planetary occ. |
| 83 | 1397 | 16″ | C.T.I.O. | pe. | intrinsic variable |
| 83 | 1510 | 8″ | Landis | pe | eclipsing binary |
| 83 | 1514 | 8″ | Landis | pe | eclipsing binary |
| 83 | 1646 | 16″ | K.P.N.O. | pe | visual binaries |
| 84 | 261 | 8″ | Cornell | pe | asteroid occ. |
| 84 | 417 | 16″ | K.P.N.O. | pe | eclipsing binary |
| 84 | 831 | 16″ | K.P.N.O. | pe | intrinsic variable |
| 84 | 1557 | 16″ | C.T.I.O. | pe | stellar color |

As is well-known to deep sky observers and photographers, it is a great advantage to observe on moonless nights in clear air well away from sources of light pollution. Less well-known is the fact that two major forms of astronomical measurements are much less demanding in their requirements. These are spectroscopy and photoelectric photometry. Photoelectric measurements of brighter stars are routinely made in the presence of the full moon from observatories located within or adjacent to major cities. Photoelectric photometry is most unforgiving when it comes to clouds, however, particularly high, thin cirrus clouds. Most small observatories cannot afford to be open every night, however, as in many cases their proprietors have to work other jobs for a living; and if there are 50 to 60 clear (photometric) nights available per year, this is all that can be realistically accommodated in many instances.

As is also well-known, larger telescopes can see fainter stars. What is not so well-known is that there are many bright variable stars of considerable scientific interest. There are many hundred such stars brighter than 8th magnitude and, by the time one reaches 11th magnitude, the number of variable stars of scientific interest are in the thousands. There are some important consequences of this with respect to telescope size and observatory location. As the telescope employed for observations becomes smaller and the background light increases (due to the moon or city lights), the effect on the photoelectric photometry is to reduce the number of the stars on which accurate photoelectric measurements can be made. It is important to note that small telescopes located in or near a city can make just as accurate measurements as large telescopes located at a dark site when it comes to the brighter stars. In fact, when larger telescopes are used to observe brighter stars, provisions must be made to "throw away" most of the light by inserting a neutral density filter or stopping the telescope down (i.e., converting a larger telescope to a smaller one). What this means is that, if you have a small telescope (6–8 inches) and operate on moonlit nights or from a city, you will have to content yourself with observing a few hundred variable stars. This is not much of a handicap, as it is hard to do real justice to even a dozen variable stars in a lifetime of observing.

What is really remarkable is that amateur astronomers equipped with small telescopes and photoelectric photometers have made and will continue to make observations from their backyards that are a direct contribution to the advancement of astronomical science. These observations are in great demand by professional astronomers, and the results are published in the most prestigious of the astronomical journals on a regular basis. While there may be other areas in astronomy and science where part-time amateurs with a minimal investment in equipment can make such a direct, needed, and recognized contribution right alongside professional scientists, astronomical photoelectric photometry is one of the most accessible and enjoyable avenues still open to the serious amateur scientist.

While many of the smaller observatories have been built and operated by amateur astronomers, there are also smaller colleges, universities, and even high schools that are equipped with small observatories. While a few of these schools have astronomy majors and full-time staff members devoted to instruction and research in astronomy, most of these schools have a more limited number of courses in astronomy and the staff has many additional duties not directly related to astronomy. These smaller school observatories are usually located on the school campus and operate under conditions not unlike those of the amateur astronomer living in or near a city. Many of the staff members at small school observatories would like to use their observatory for serious scientific research for two reasons. First, engaging in serious research is very instructive for students being trained in the sciences. Second, the staff at these schools have received training in astronomy or related sciences (often at the graduate level) and they rightly desire to exercise their professional capabilities not only by teaching, but also by engaging in productive research. Photoelectric photometry is particularly amenable to the needs of small observatories at schools, and the observatory and time of staff and students can be used productively in conducting serious astronomical research that results in the publication of results in the professional journals.

## 1.3   Recent Advances in Photoelectric Photometry

Until recently, the number of small school and amateur observatories regularly engaged in photoelectric photometry was not large. There were a number of reasons for this. First, and perhaps most significant, it was simply not known very widely that small observatories could make these types of observations, that the observations were really needed, and that the professional journals desired to publish such observations. Of almost equal importance, until recently photoelectric equipment tended to be both expensive and finicky and its operation required a familiarity with specialized electronics. Finally, the extensive data analysis required after the observations were made was difficult to understand and tedious in execution. The result of all this was that the great bulk of photoelectric photometry was done on the small telescopes at the large observatories by professional astronomers.

Most of the information needed to do photoelectric photometry is now available in convenient book form. While outdated somewhat, the two books by Wood (1953 and 1963) still provide much valuable reference material, as well as an appreciation of the earlier approaches to photometry. Young (1974) gave a detailed treatment of photometry that remains a primary reference well worth studying. The book by Golay (1974) contains a detailed mathematical treatment, while the first edition of this book (Hall and Genet, 1982) contained many practical details. The book by Henden and Kaitchuck (1982) covers all aspects of astronomical photometry and is a classic that should be studied by every photometrist. The two-book series *Advances In Photoelectric Photometry* (Wolpert and Genet, 1983 and 1984) contains many useful suggestions relative to small-observatory photometry, while another two-book series, *Microcomputers In Astronomy* (Genet 1983b, and Genet and Genet 1984), is primarily devoted to photometry. There is a book entirely devoted to software for photometry, *Software for Photoelectric Photometry* (Ghedini, 1982), and another, *Solar System Photometry Handbook* (Genet 1983c), is dedicated to solar system photometry. Major sections of *Microcomputer Control of Telescopes* (Trueblood and Genet, 1985) are devoted to photometry, particularly automatic photometry. The photometrists in New Zealand have produced two book-length special issues of *Southern Stars*, dedicated entirely to photoelectric photometry at smaller observatories, that contain many valuable chapters well worth reading. Percy (1986) organized a symposium devoted entirely to the study of variable stars with small telescopes containing excellent overview chapters on numerous aspects of photometry. NASA sponsored a symposium on the topic of improving the accuracy of photometry and published the book-length proceedings (Borucki and Young, 1984). An IAPPP symposium on automatic photoelectric telescopes also produced book-length proceedings (Hall, Genet, and Thurston, 1986), as did an IAU symposium on instrumentation and research programs for small telescopes (Hearnshaw and Cottrell, 1986).

Besides the increase in readily available information, photoelectric equipment has greatly benefited from the revolution in solid state electronics over the past decade. No longer is the photoelectric equipment expensive and finicky. Highly reliable, easily operated photometers are now available at moderate prices. The photometers available commercially from Optec, Inc. and EMI Gencom are prime examples of such modern equipment. The electronics revolution has also benefited those who desire to build their own photometers, greatly simplifying the construction and improving performance.

Finally, the ready availability at low cost of programmable calculators and microcomputers has turned the once formidable task of data reduction and analysis into what is only a minor irritant to some and a downright pleasure to others. While many excellent observations continue to be made without the assistance of microcomputers in the data reduction and analysis, the trend is clearly towards greater use of microcomputers.

## 1.4    The Future of Photoelectric Photometry at Small Observatories

The past five years have seen a very large increase in the number of small observatories around the world equipped for photoelectric photometry. There has been an even greater increase in the number of full-time professional astronomers who have recognized the usefulness of asking small observatories to make photoelectric observations. We expect that these two trends will continue for some time. Moreover, the result is the almost paradoxical situation that, as the number of photoelectric observations made by small observatories increases, the number of unfilled requests by full-time professional astronomers increases even faster.

As reliable, low-cost photoelectric equipment has become available, it has allowed an increase in the number of small observatories equipped for photoelectric photometry. This, in turn, has stimulated the demand for reliable, low-cost photometers which has resulted in new entries in the market place. If this trend continues, it is bound to result in photometers that perform even better than those available today, are more reliable and easy to operate, and are even lower in cost. The surge of interest in small observatory systems in operation has also stimulated the design of equipment that can be easily constructed by the proprietors of small observatories. Many of the amateur astronomers now involved in photoelectric photometry are full-time electrical engineers or physicists by profession; they in turn are bringing their knowledge of the latest advances in electronics and computers to bear on devising better equipment.

Another trend that is already visible and should be expected to continue is increased cooperation between amateur and professional astronomers. While most professional astronomers have always been rather kindly disposed towards amateurs, it is understandable that professional astronomers might have a less than keen interest in the photographing of deep sky objects with small telescopes. On the other hand, the use of small telescopes for photoelectric photometry is something that they themselves do, and in fact is the primary role that they feel smaller telescopes should play in serious research. Thus, photoelectric photometry with smaller telescopes not only provides a common ground of interest between the amateur and professional astronomer, but it allows the amateur and small school professor to make the same measurements the full-time professional would have made if he had the time to do it.

As mentioned earlier, most of the larger observatories have one or more smaller telescopes which are devoted almost entirely to photoelectric photometry. Kitt Peak National Observatory had, until very recently, two 16-inch telescopes used almost entirely for photoelectric photometry, and there are similar telescopes at Cerro Tololo, etc. What was not mentioned earlier was that these telescopes are in great demand and rather heavily booked. A result of this is that a professional astronomer requesting time on one of these telescopes is usually limited to one or two weeks per year. As might be expected, this tends to encourage observing projects that can be completed in this amount of time (as publication of results is expected if one wants to be invited to return) and tends to discourage observing projects that are long-term in nature. This is not to suggest that these smaller telescopes at the major observatories are not productive: in fact they are the most productive, in terms of published research, of any telescopes in the world. The point, however, is that the amateur or small school observatory is available to the proprietor all year round rather than for just a week or two per year. This means that longer term projects such as photoelectric observations of long-period variables are possible. While full-time professional astronomers would like to be able to take on longer term projects and they feel the need for such research, the fact of the matter is that in general they simply do not have either the observing time or telescope availability to take on such long-term photoelectric observations. The result is that this is one area of serious astronomical research where the amateur and small school observatory have an advantage. This is especially so if these small observatories equipped for photoelectric photometry organize themselves and work in concert. With their potential for large numbers and capability of international organization, the day (well, night) is just around the corner when small observatories will be able to keep hundreds of stars under almost constant photoelectric surveillance around the clock, providing measurements of the best accuracy attainable by any means in a manner essentially identical to that done by full-time professional astronomers.

## 1.5   International Amateur–Professional Photoelectric Photometry

On a professional level, Commission 25 of the International Astronomical Union (IAU) was formed a number of years ago to coordinate professional activities in the area of photoelectric photometry and polarimetry (the latter is closely related). IAU Commission 25 has done an excellent job in coordinating professional activities in this area. There have also been a number of regional organizations that have done much to promote cooperation in the area of photoelectric photometry. The Royal Astronomical Society of New Zealand (RASNZ) and the American Association of Variable Star Observers Photoelectric Committee (AAVSO PEP Committee) are good examples of organizations that have done much to develop the use of photoelectric photometry by small observatories in their regions. One of us (DSH) has made consistent use of amateur and small school photoelectric observations in professional research for over a decade. With the recent surge of photoelectric capabilities in small observatories around the world, there has been an increase in the demand for readily understandable information on photoelectric photometry. The International Amateur-Professional Photoelectric Photometry (IAPPP) association was formed in June 1980 to meet these needs. It does this through a quarterly journal, the *IAPPP Communications*, and through annual IAPPP Symposia held at various places throughout the world to afford a direct face-to-face exchange between amateur and professional astronomers engaged in photoelectric measurements. From its very inception, IAPPP has been viewed as an equal partnership between amateurs and professionals, and the current membership of over 800 from more than 40 countries reflects this spirit in that the members are almost evenly split between amateur and professional astronomers.

IAPPP's role is intended to be solely supplemental and supportive to that of other organizations and established journals, and it achieves this by concentrating on the practical details of equipment, data reduction, computer software, observational techniques, and observing programs not normally covered in the regular journals and symposia. IAPPP does not publish observational data or results as such, because the ultimate goal of IAPPP is to see reliable and useful photoelectric photometry from smaller observatories appear in reputable astronomical publications where it will be accessible to the professional astronomical community.

Both amateur and professional astronomers involved in, almost ready for, or just interested in photoelectric photometry are invited to join IAPPP. Annual dues are $15.00 U.S. and a check or money order payable to IAPPP should be sent to *IAPPP Communications* Assistant Editor, Robert C. Reisenweber. All members receive the quarterly publication *IAPPP Communications*. Contributions for publication are solicited from members in the following areas:

1. Suggested observing programs of current scientific interest to professional astronomers that are suitable for photoelectric photometry at smaller observatories.

2. Equipment design and construction.

3. Observational and data reduction techniques.

4. Descriptions of smaller observatories equipped for photoelectric photometry.

5. Reviews of recent meetings, articles, and equipment relevant to photoelectric photometry.

6. Announcements of meetings or other matters relevant to IAPPP.

7. IAPPP Committee reports.

8. Letters to the Editors.

9. References to published photoelectric observational results by smaller observatories.

10. History of photoelectric photometry.

IAPPP sponsors and encourages both informal regional gatherings and more formal symposia. The symposia are usually held jointly with other societies and concurrent with or adjacent to other meetings.

The many back issues of the IAPPP Communications provide perhaps the most valuable information of all for the new photometrist. Virtually all topics of interest to photometrists at small observatories have been treated in detail, and one can get a better feel for the entire field by reading these back issues than perhaps from any other source. Back issues are available from IAPPP itself and also from Willmann-Bell, Inc., the publisher of this present book. We cannot recommend careful study of these back issues too highly.

## 1.6   Overview of the Remainder of the Book

In spite of improvements made to equipment in recent years, the fact remains that photoelectric photometry is a highly specialized science that requires specialized knowledge, skills, and equipment. Making good observations requires considerable patience, persistence, and care. No one should expect serious contributions to science to be easy, and photoelectric photometry is no exception.

What we have tried to do in this book is to cover all the things we feel you need to know to get started in photoelectric photometry. To become really proficient in this area will require not only what is in this book but also a good dose of actual hands-on experience in making observations and further reading and study from the various references provided. If this is done and if one mixes some face-to-face discussions with other amateur and professional astronomers working in this area, it is entirely within the capability of almost any amateur astronomer to become a proficient photometrist.

Chapter 2 traces the history of photoelectric photometry from its beginnings just before the turn of the century (by an amateur), through the first methodical observations by Joel Stebbins, and on to the subsequent advances that turned photoelectric photometry into a major area of modern astronomy. This exciting history is appropriately related by Gerald E. Kron, a student of Joel Stebbins. Kron was the first to use a photomultiplier in astronomical photometry and it was this, more than anything else, that caused the sudden spread of photoelectric photometry to observatories around the world immediately after World War II.

Chapter 3 begins the equipment section of the book, starting with the observatory itself. It suggests that the best location for an amateur observatory is in the backyard and that the best location for a school observatory is on campus. As useful photometry requires a reasonably regular schedule of somewhat tedious and long observations at the telescope, suggestions are given for making the observatory as comfortable and convenient as possible. Chapter 3 also discusses the telescope. It suggests that, if you already have a telescope, this is the best one to start with. For those planning to acquire or build a telescope from scratch or for those considering making improvements on an existing one, this chapter suggests the features in a telescope that have proven to be most critical when used for photoelectric photometry. Important telescope accessories such as the finder, setting circles, and fine adjustments in right ascension and declination are also discussed.

Chapter 4 covers the photometer head. After introducing the overall functions and makeup of photometer heads, the individual components of the head are discussed separately. These include the previewer, diaphragm, microscope, filters, Fabry lens, photodetector, and cold box. Physical considerations such as attachment to the telescope, avoiding light leaks, alignment, and ease of operation are covered. Considerations that should be made in selecting a photometer head are given. Detailed examples of actual photometer heads are then presented, including both photomultiplier-based systems and those using a photodiode. The examples also include commercially manufactured units currently available on the market, as well as units designed for ease of constructing them yourself.

Chapter 5 covers the electronics required to make a photometer work, including high-voltage power supplies, DC amplifiers, and pulse amplifier/discriminators. A number of examples of each type of support electronics are given. There is also a discussion of the relative merits of DC amplification vs. pulse counting as it applies to photoelectric photometry at small observatories. Chapter 5 also discusses data recorders. These include meters, strip chart recorders, and counters.

Chapters 2 through 5 cover equipment basics in photometry. Chapters 6 and 7 cover two advanced topics in photoelectric photometry: computerized data logging and automatic photoelectric telescopes. For the beginner, these chapters should be omitted, and this may be done without any loss of continuity whatsoever. It should be noted that many amateur and professional observatories do quite well indeed without any computerization whatsoever. On the other hand, advances in electronics are making some degree of automation increasingly attractive for advanced practitioners.

Chapter 8 concludes the equipment section of the book with a description of a number of smaller observatories equipped for photoelectric observations of variable stars. While the features of particular interest of each observatory are described, the photographs, which were kindly provided by the individual observatories, are the main message. Our only regret is that space precluded more than a mention of the dozens of other outstanding small observatories equipped for photoelectric photometry.

Chapter 9 begins the nonequipment section of the book. It discusses the earth's atmosphere. Atmospheric extinction, dispersion, scintillation, and sky brightness are all discussed in detail. Emphasis is placed on how these affect photoelectric photometry.

Chapter 10 discusses standard photometric systems with emphasis placed on the UBV system. Standard photometric systems are required so that observations made at different observatories using different equipment can be meaningfully combined. The determination and use of transformation coefficients to translate results from a local instrumental system to a standard system are covered in considerable detail.

Chapter 11 then covers differential photometry, the approach to photometry most frequently used by most smaller observatories. The basis and mathematical formulation of differential photometry is treated in some detail.

Chapter 12 discusses observing techniques. Included are suggestions for choosing a comparison star, monitoring the sky, making a deflection, a recommended basic sequence of readings for differential photometry and reasons for altering this sequence, multi-filter differential photometry, choice of diaphragm size, recording the time, what to write down during observing, determining extinction, and determining transformation coefficients.

Data reduction is covered in Chapter 13. Included are instructions for determining raw instrumental magnitudes from deflections, determining instrumental differential magnitudes, correcting these for differential atmospheric extinction, transformation of the results to the standard system, making the heliocentric correction to the time, and the computation of phase for eclipsing and other periodic variables.

Chapter 14 concerns achieving maximum accuracy. No fewer than 22 suggestions for improving accuracy are given. These suggestions are based on years of experience and include all the problems that have tended to hurt the accuracy of the observations made by beginners in photoelectric photometry. While there is no substitute for experience, these suggestions will put you on guard against the many possible ways in which your observations could be corrupted. This chapter also discusses how one can calculate the standard deviation and the mean error of a sequence of photoelectric observations.

Chapter 15 discusses some of the factors relevant in the choice of an observing program. Programs include long-period variables, short-period variables, and extremely rapid variables. A discussion of programs where wavelength is and is not important is given. Programs requiring solid versus intermittent coverage are also presented. Finally, situations when your particular geographical longitude and latitude can make you valuable are discussed.

Chapter 16 suggests how to select an observing program. These suggestions include observing programs recommended by another astronomer, requests for observations in the *IAPPP Communications*, times of minima of eclipsing binaries, discovering new variables, lunar occultations, photoelectric sequences, and RS Canum Venaticorum binaries.

While this book is intended primarily for the beginner in photoelectric photometry, we hope that it will provide a few new thoughts to the advanced practitioner as well.

# Chapter 2

# A SEARCH FOR OUR ROOTS

## 2.1 Prologue

The problem of tracing the history of something—anything—depends a lot upon the nature of the "thing." Take the sweet potato, for example; its modern history is simple. We know where it is found and used, and we know how it is prepared for food all over the world. Its origin, however, is something quite different. How can we study the origin of something that may have been on this planet longer than we have? Clearly, the problems involved here are concentrated in the past, and are well described by the impressive cliché "lost in the dim reaches of the past."

If one wishes to study the history of a modern technique—say astronomical photoelectric photometry—the past will be simple, well-documented, *and* not very remote. The past of electrical photometry cannot reach back in time prior to the beginnings of the practical use of electricity. Here, though, the *present* is something quite different; for it we can invent a neat new cliché: "lost in the confusing jungle of modern technical development."

Of course, we will start with the early days and let the complicated jungle phase go until later. In what follows we will pay attention only to results, and will ignore unimplemented ideas. Leonardo da Vinci invented a flying machine, but we had to wait for the Wright brothers to create a functional self-powered man-carrying machine.

## 2.2 Earliest History

The first documented photoelectric photometry was done in Great Britain by four men: G. M. Minchin, M.A., F.R.S., Professor of Mathematics; Mr. Monck of Dublin, possessor of a 9–inch refractor; Professor Fitzgerald; and Mr. W. E. Wilson, possessor of a 24–inch reflector. Monck and Wilson must have been English amateur astronomers, Fitzgerald probably had the quadrant electrometer, and Minchin (1895, 1896) made the selenium photocells and published the results. The very first work was done by Monck and Fitzgerald in Dublin, employing a Minchin selenium cell in 1892. With a 9–inch refractor, Monck and Fitzgerald were able to get appreciable electrometer deflections from the light of Jupiter, Venus, and possibly Mars. This accomplishment is documented only as a statement at the end of Minchin's 1895 paper.

Between 1892 and the observations made in April 1895, Minchin improved his selenium cell. The improved cell, used with Wilson's 24–inch telescope, made possible quantitative measurements of the light of several bright stars, a feat impossible for Monck and Fitzgerald. One can visualize the scene. Wilson and Minchin are in the dome of the 24–inch, Fitzgerald down below the floor with the electrometer. Wilson sets his telescope on Arcturus; Minchin centers the image and shouts to Fitzgerald to be ready; Minchin slides the cell aperture into registration with the image, and the two men at the telescope wait eagerly for word of what happened. Eight millimeters! A deflection of 8 millimeters! Stellar photoelectric photometry had become a fact.

It would be a disservice to Minchin if I did not describe his ingenious selenium cell, especially since it embodies a principle the value of which has been grasped again only relatively recently by astronomers. Selenium reacts to illumination in two different ways: it can generate a voltage and it can change its conductivity. A feature of Minchin's cell made it necessary for our four pioneering gentlemen to measure the voltage produced by the selenium upon its illumination at the telescope. This they did with their quadrant electrometer, an instrument intended mainly for the measurement of electric charge, but capable of indicating the application of a voltage.

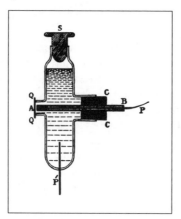

**Figure 2–1.** *Cross-section of the Minchin selenium cell.*

Minchin's selenium cell is shown in cross-section in Fig. 2–1; the cell is small, only 3 cm long and 1 cm in diameter. QQ is a quartz window cemented to the glass body; behind the window at A is the tiny sensitive surface, photoactive selenium deposited on an aluminum cylinder held in the glass tube AB. P and P′ are platinum wires sealed into the glass and serving as the two electrodes. S and CC are stoppers that seal into the container an electrically conducting liquid, oenanthol. According to the *Merck Index*, oenanthol is n-Heptyl alcohol; Minchin does not say why this particular liquid was chosen. The liquid serves to make electrical contact with the front surface of the selenium, the electrode connected to P to the rear surface. Why is the selenium surface so small? This feature can be grasped by a reduction to absurdity: What if it were a foot in diameter? Better yet, hear Minchin's own words on the subject: "The seat of the electromotive force is the surface of contact of the liquid and the selenium ... If, now, $P$ is connected to one pole of an electrometer and $P'$ with the other, and if there is any portion of the selenium surface which is not exposed to light (and consequently not the seat of an E.M.F.), this inert portion will act simply as a conductor conveying a portion of the charge to the wrong pole of the electrometer, and thus giving a diminished effect." More modern photo-conductive materials such as cadmium sulfide and lead sulfide were also made in small sizes and for the same reason. One lead sulfide cell made by Eastman Kodak was a square 0.5 mm on a side. A practical problem arises in keeping a star image accurately centered on such a small area. Minchin *et al.* were the first astronomers to experience this problem, as brought out by them in the 1895 paper.

Minchin's 1896 paper was very short, only $2\frac{1}{2}$ pages. In it our three gentlemen reported on improvement in the selenium cell, and on measurements of the radiant flux from ten bright stars. Some calculations were made of the total power emitted by some of the stars, and further work of the same kind was promised. However, this paper was the last word. One wonders why; one also wonders about the fate of the 24-inch telescope, of the electrometer, and of the selenium cells. Are they all still there in that English country house?

## 2.3 The Stebbins Era

Professor Joel Stebbins furnished continuity to the development of photoelectric techniques for 50 years. Stebbins was not adept at technical development; he always depended on others for that aspect of his work. However, he had a full appreciation of technical methods, and he spent most of his life coaxing technically capable people to cooperate with him in implementing his enormous zeal for astronomical research. Stebbins contrived to apply the selenium cell to photometry prior to learning of the work of Minchin, and in 1913 he was one of three persons who independently conceived of employing the photoemissive cell for the same purpose.

Stebbins (1940) began working with the selenium cell in 1907. Stebbins and his technical colleagues measured the change in resistance of selenium when exposed to light. For this they employed a Wheatstone bridge circuit with readout by galvanometer. Improvements were made on the original equipment until bright stars could be measured with precision limited by atmospheric conditions. One of the improvements coming from this work was the use of refrigeration; an ice pack around the cell holder considerably improved the signal-to-noise ratio, and thus Stebbins became the first to use refrigeration. He told me that they got the clue to refrigeration by working on naturally cold nights at Urbana, Illinois. Stebbins also told me that they dropped one of their selenium cells on the floor, that the cell broke in half, and that when they tried one of the halves, it worked better than the whole cell! This was an important clue whose significance may have been missed by Stebbins and colleagues, unless one of the "improvements" they made was to quietly make their cell still smaller by trimming more useless material from it. These cells were enormous. They were made for on-and-off control purposes by illumination from bright, nearby sources, and were not at all well adapted for detection and measurement of small, faint light sources.

**Photograph 2–1.** *Joel Stebbins in 1957, one year before he retired from astronomical work, and 50 years after he first illuminated a selenium cell with starlight. (Photograph provided by Gerald E. Kron.)*

In 1913, Dr. Jakob Kunz began furnishing potassium hydride photoemissive cells to Stebbins, who started using them at once. In the same year Guthnick in Berlin and Rosenberg in Tübingen undertook similar work, but only Stebbins carried on and on, first at Urbana, and after 1922 at Madison, Wisconsin. The parallel activity in Germany during those decades is discussed by Walter (1985).

The photoemissive cell is a device that emits a stream of electrons upon being illuminated; that is, it provides an electrostatic charge, a charge that can be accumulated by a capacitive device until the rising voltage becomes high enough to be read out with desired precision. The electrometer is a device that provides both the accumulative capacity and the voltage measuring ability in one instrument. However, compared with the properties of single electrons, practical electrometers that can be carried on a moving telescope are clumsy things with characteristics that would require the observer to make rather long exposures in order to collect enough charge for the signal-to-noise ratio to be set by the photocell. Long exposures require guiding, and thus an observer with primitive telescopic equipment generally cannot realize in practice his theoretical limiting magnitude. This situation is considerably relieved by placing an inert gas in the photocell which ionizes from bombardment with the photoelectrons and can produce a gas multiplying factor of ten or even more. A definitive discussion of the theoretical maximum performance of the simple diode photocell has been given by Smith (1932), in a paper that is required reading for anyone wishing to have a fundamental knowledge of what is going on during photoelectric photometry.

Stebbins carried on with a photocell-electrometer combination from the start in 1913 until Albert Whitford developed the electrometer thermionic amplifier in 1932. The string electrometer was used until the invention of the mechanically more stable Lindemann electrometer, which was substituted in 1927. A photometer with a Kunz cell and a string electrometer was built at the Lick Observatory and used by Cummings (1921) during the early to mid-1920's. The superior properties of the Lindemann electrometer encouraged E. A. Fath to build a photometer used by him on the Lick 12–inch refractor for many years of measurements of variable stars. The influential work of John Hall was also done with an electrometer but under unusual conditions to be described later.

It may seem to be a simple matter to "amplify" the output of a photocell with a vacuum tube, solve all of the residual problems of photometry at once, and happily enter the Elysian Fields of perfection. In the event, it took Whitford (1938) with the FP–54 electrometer tube in hand and Joel Stebbins lurking in the background acting as catalyst and entrepreneur to bring photoelectric photometry out of its dark ages. Perfection was still for the future, but in the excitement of Whitford's accomplishment this was easy to ignore. The hardware cost only a few hundred dollars and, with an apology to Albert, his salary probably did not add much more to the total cost of the development. The transformation from electrometer to electrometer tube took place very quickly. The rewards were enormous. The telescope carried the photocell, grid resistor, and electrometer tube in a brass tank. Readout was done with a galvanometer, the only mechanically sensitive element of the combination which, on the end of a pair of wires of any length, as it was, could be put anywhere, but in fact was always attached to the telescope pier. The electrical circuitry to control the whole thing was placed on a desk conveniently located with respect to the galvanometer. The astronomer generally sat at the desk reading the galvanometer, twiddling knobs, and giving orders to an assistant who operated the telescope. Although astronomer, desk, and galvanometer could have been in a warm room, astronomers were practicing masochists in those days, and for many years the two operators willingly suffered together in the cold.

One of Whitford's accomplishments usually overlooked was to isolate the electrically sensitive part of the photometer in a sealed metal tank from which the air was evacuated. This procedure solved the problem of the effects of moisture on critical electrical insulators, and enabled the photometer to be used on damp nights, to say nothing of causing a general improvement of noise level. Whitford built two photometers according to his design, one being installed on the 15–inch refractor of the Washburn Observatory in 1932. The other was taken to the Mount Wilson Observatory where it was used for many years on both major telescopes by Whitford and Stebbins in the latter's capacity of Research Associate.

**Photograph 2–2.** *John Hall, Gerald Kron, Joel Stebbins, Albert Whitford, and Canadian astronomer Beals in 1939. Photograph taken by John Hall with a delayed shutter.*

The way now seemed clear for Stebbins and Whitford (and others who began to enter the field) to relax, forget instrumental development and get on with doing astronomy. However, in the same critical year, 1932, John Hall further stimulated technical development with his important accomplishment of bringing the Cesium-Oxide-on-Silver photocell into astronomical service (Hall, 1932). The Cs–O–Ag photosurface was by no means new; it is an excellent device with the virtue of good efficiency over a very wide range of wavelengths, but with the serious vice of giving such a high spontaneous emission of thermal electrons as to render it useless for the measurement of faint light. Hall and the Bell Labs scientist, C. H. Prescott, Jr., collaborated on the construction of a fine, double-ended photocell that was later manufactured for sale by Western Electric under the name of the D–97087. Hall experimented with cooling one of these cells at the suggestion of Prescott and found that, at the temperature of dry ice, the dark current all but vanished. Hall was then able to construct a useful photometer employing dry-ice cooling. The technical problems were alleviated for Hall by his access to the Loomis telescope at Yale University. This telescope was fixed, with the light being fed to the objective by means of a coelostat, and the problem of fitting a cooled cell, closely coupled to an electrometer became tractable. Hall went on to build another photometer and produce photoelectric spectrophotometry over the wavelength range of the D–97087 for a significant number of bright stars.

Stebbins and Whitford quickly adopted Hall's development and Whitford soon produced a refrigerated photocell using his amplifier. The Stebbins and Whitford (1943) six-color photometry was devised and used by them at Mount Wilson until the onset of the war, and by Stebbins and myself after Stebbins' retirement in California after the war, until Stebbins had to stop working in 1958. After this, I carried the six-color work on, with the collaboration of others until my retirement in 1973.

Refrigerated photometers opened up new fields in astronomy, but refrigeration also opened up new fields in technical problems. Whitford wanted to continue using evacuation to solve moisture problems. At that time, around 1934, vacuum technology was in the sealing-wax-and-stopcock-grease phase of development. The photocell and amplifying tube were both rather large pieces of equipment, and so only the cell was cooled. The glass container of the cell was attached to the metal container of the tube by means of a waxed joint which often, because of embrittlement from at least winter ambient cold, if not from the dry ice, would crack and leak. The moisture and resulting internal fogging could bring work to a sudden end. Then, there was the window. The window got cold, too, and tended to fog, so it was heated with an electric heater, but this increased the use of dry ice and also warmed the cell to some extent—and so on and so on.

After the war, Whitford and I continued to grapple with these problems, and our collaboration resulted in a funny incident. I had tried a cellophane window on the theory that the radial thermal conduction of such a thin layer of cellulose would be so small that the window would remain above the dew point. An experimental window survived the stress of evacuation and seemed to work during several nights of observing, so I told Whitford about it and he tried the scheme at Mount Wilson. Shortly after, I received a telegram; at Mount Hamilton these were always telephoned and then mailed to us later. The girl on the phone said: "You have a telegram, sir, but it consists of only two words, and as I can't understand it, I think that it must be garbled. I'll read it to you anyway; it says: 'Cellophane, BAH!'." Window fogging wasn't the worst, either. Cemented windows would spring leaks, and on cold nights the whole photometer would sometimes cool below the dew point and even the filters would fog, to say nothing of the rest of the optical system. After I moved to the U.S. Naval Observatory in Flagstaff in 1965, we solved these problems permanently. The window was made of clear sapphire brazed into a thin-walled stack made of Kovar, a metal of low thermal conductivity. The window end of the stack was in thermal contact with the rest of the photometer, *all* of which was gently heated. No more fogging of anything, no more sticky lubricants, and no more cold hands.

The invention of the photomultiplier did not quite end the use of the diode photocell. Although the new alloy photosurfaces were much superior to the potassium hydride surface, they were not sensitive to the near infrared, and so were not suitable for the six-color photometry. After his retirement, Stebbins came to work at the Lick Observatory where he and I collaborated doing six-color photometry. I was unwilling to use our old infrared photometer because of its mechanical unreliability. During the war a new electrometer tube appeared on the market along with a new kind of high resistance resistor, both made by the Victoreen Company. The tube was very small. Excellent small Cs–O–Ag gas-filled photocells were made by Continental Electric. The three small parts, tube, cell, and resistor, could now reside together in a small tank only 3 inches in diameter by 3 inches long. The technical problems of refrigeration were greatly reduced, grid resistor noise was slightly reduced, and reliability was much increased. With a further minor improvement of substituting a slush of dry ice in isopropyl alcohol, Stebbins and I used this equipment until the availability of Cs–O–Ag photomultipliers put all of this development (Kron, 1947) out of business.

The small size of components encouraged one more (and final) development of diode photocell photometry by Kron and Fellgett (1955). The Victoreen 5800 electrometer tube seemed to work better when cooled (lower grid current) and not to suffer in any way, whereas grid resistors sometimes became troublesome when cooled and had to be selected. When cooled with liquid air, grid resistors became inoperable. Peter Fellgett and I designed and built a photometric receiver that eliminated the grid resistor by reverting to the rate-of-charge method, which performed extremely well. Refrigeration with liquid air was not ruled out by anything we knew, and would have further improved performance. This development, too, as well as the Stebbins era itself, was brought to an end by the photomultiplier.

## 2.4   The Photomultiplier Era

Although the photomultiplier was invented in the mid-1930's, the development needed to bring it into production along with the intervention of the war took so much time that photomultipliers did not begin to be used by astronomers until 1946. There is no need to describe a photomultiplier here and neither is there a need to define what one is. See, for example, Kron (1981). The introduction of this device greatly simplified the course of instrumenting photoelectric equipment. Along with the introduction of the photomultiplier came many other profound changes caused partly by a much greater level of funding in the sciences, but also by the development, during the war, of technically trained people, an awakened interest in the sciences, and the availability of wartime technical advances. Often overlooked is the importance of the new alloy photocathodes discovered in the early 1930's. These cathodes are easy to make, withstand abuse, have quite a long color-sensitivity range, have high quantum efficiency, and serve well as secondary-emission surfaces in photomultipliers. Had the potassium hydride and Cs–O–Ag been the only photosurfaces available for use in multipliers, these devices would not have achieved the importance that they have. The convenience of the photomultiplier led quickly to the establishment of the extremely important UBV system of photometry by H. L. Johnson and W. W. Morgan 30 years ago, and still in use today. High-gain D.C. amplifiers were developed quickly, and strip chart recorders were soon put into service to give and preserve a complete history of the photometer output. Successful photomultipliers having Cs–O–Ag photocathodes were developed by André Lallemand by making the secondary emitters of a material different from that of the cathode. By the mid-1950's many astronomers were able to settle down to the routine use of simple, accurate, and inexpensive photometers. Photoelectric photometry had come of age and was accepted in all observatories as an important and permanently useful tool. I am glad that Joel Stebbins lived to see the fruition of this era in astronomy.

Further development has been valuable but not of fundamental importance. There are new photocathodes, some very expensive, all with unusually high quantum efficiency. There is much automation; photometers can now be operated from a remote station, and the output can be recorded on magnetic tape and processed later by computer. One wonders if the astronomer is not being isolated from needed intimate contact with his data. Stebbins would not have liked it. But enough of this.

Shortly after the war Whitford engaged in an intense development of cadmium and lead sulfide infrared-sensitive photoconductive cells, and with this we have come a complete circle from 1895. Others have carried on with fantastic detectors requiring cooling with liquid helium, and enabling photometry at wavelengths of many tens of microns. The developments of the last 15 years are overwhelming, and we find ourselves in the jungle. It will take someone better informed that myself to write a history of this period.

## 2.5  Panoramic Photometry

Panoramic photometry has been done by electronic as well as photographic methods for more than 20 years. With aperture or diaphragm photometry a single object is observed at a time and the data accumulated first and recorded later. The simplest and cheapest method of recording is to write down a number. In panoramic photometry the recording is done first (for example, a picture is formed) and the data are extracted at any appropriate time later for any number of the many objects that may have been recorded.

The jungle contains many kinds of instrumentation that seem to have photometric potential. At present, however, the most promising are the charge-coupled device and the electronic camera. Both are extremely expensive. The very large quantities of data yielded can be handled only by a computer of considerable capability. It appears unlikely that panoramic photoelectric photometric systems will ever be of interest to any but very wealthy and very courageous amateur astronomers. However, their appearance is part of our history, and so it is appropriate to devote a few lines to them.

The charge-coupled device is a silicon chip on which has been formed a two-dimensional array of photodiodes. The best available ones (1982) are $800 \times 800$ arrays of tiny square diodes 15 microns on a side; thus, the array contains 640,000 pixels in a square that is 12 mm on each side. When illuminated, each diode accumulates an electrostatic charge which is drained off periodically from two edges. The digital output resulting from this process is stored by means of magnetic mass storage. Storage capacity is unlimited, as the chip can be charged and dumped as many times as the operator wishes for a given field. The data can be integrated later by a computer and processed as one wishes, including the production of a synthetic picture of the field. Silicon diodes can have quantum efficiencies of up to 60% and they have sensitivity over a wide color range; it is easy to see that charge-coupled devices have great potential, though we need larger and more nearly perfect chips.

The first functional electronic camera was produced by Lallemand (1936) who went on to produce two practical versions of his classic camera after the war. A review of electronic camera development up to 1974, along with construction details, has been given by Kron (1974). Since 1974 two excellent cameras with 8–cm diameter cathodes have been developed, one by D. McMullan (Royal Greenwich Observatory) and one by Paul Griboval (University of Texas).

**Photograph 2–3.** *The Lallemand electronic camera on the nebular coudé spectrograph of the 120–inch telescope. Lick Observatory. (Left to right) R. M. Duchesne (collaborator with Lallemand), G. E. Kron, A. Lallemand, and M. F. Walker.*

It is interesting to note that electronic cameras have generally been developed by astronomers and built in observatories. Thus, the electronic camera is not in production and cannot be purchased on the market, the main problem with this otherwise practically perfect photometric instrument.

The electronic camera produces a picture on silver halide emulsion that looks just like a photographic negative. However, the emulsion is exposed to electrons, not photons, and hence the picture is an electrograph, not a photograph. The electronic camera separates the functions of detector and storage; detection is done by a semi-transparent photoelectric cathode, storage takes place in a special very fine grained silver halide emulsion. The cathode can be used indefinitely. Each photoelectron, focused electronically upon the emulsion, creates from five to ten developable grains, so that single electrons are easily detectable. The emulsion contains so many grains that loss of sensitivity from grain exhaustion is trivial within the density limit (about four) imposed by microdensitometers. Useful pictures can be made with the electronic camera alone, but photometry depends upon measuring emulsion density with an accurate scanning microphotometer, such as the PDS. The "pictures" constitute permanent storage of very large quantities of photometric data in a form that can be qualitatively examined or accurately measured at any time. In the form of the pictures, enormous quantities of data can be stored in a trivial space, such as a small card filing cabinet. The Lallemand electronic camera is shown in Photo. 2–3.

## 2.6  Epilogue

This history is no doubt quite incomplete. This I admit. It was written from the point of view of one who has lived and worked throughout the time interval during which the development of modern photoelectric photometry took place. In order to keep the story to a reasonable length, I have had to omit contributions from many workers and slight the contributions of others. For this I apologize, but there was no apparent alternative. The point of view has been more that of a technician than that of an astronomer; a valuable contribution could be made by someone else writing from, so to speak, the outside looking in.

**Photograph 2–4.** G. E. Kron at his Pinecrest Observatory in Flagstaff in 1981. He is surrounded here by the computer and tape drive he uses to reduce data from satellite observations. (Photograph by J. L. Hopkins)

# Chapter 3

# THE OBSERVATORY AND TELESCOPE

## 3.1  Introduction

The observatory which houses the telescope and photometric equipment is an important element of the total photometric system. Compared to other astronomical activities, photoelectric photometry places requirements on the observatory which in some cases are more demanding, while in others, surprisingly, are somewhat more relaxed. With these requirements in mind we must discuss the observatory's location and physical structure.

Much successful photoelectric photometry has been conducted without an observatory as such. Some telescopes, such as the Celestron C–8, are quite portable and can be set up quickly. With special care, the photometer head and electronics also can be set up quickly. For many situations, this is the best approach and should be given serious consideration.

## 3.2  Location

The location of the observatory, when one is free to choose, is an important consideration. The most important factor is convenient, ready access. An observatory on a dark mountain top is of little benefit if it is hardly ever used. For the amateur the best location is the backyard and for the small college, near or on the campus. We cannot overstress the importance of a convenient location, and the utility of the backyard or campus observatory. Small observatories at remote locations simply see little use, and photometry requires sizable amounts of observing time. Photoelectric photometry is highly tolerant of background sky light, provided one is not observing stars which are too faint. One illustration of this is the fact that few photoelectric photometrists interrupt their observing programs on brighter stars even when the full moon is high in the sky.

Louis Boyd operated a 10–inch automatic photoelectric telescope (APT) from essentially downtown Phoenix for several years with excellent results. The measured city sky brightness was one-third that of the full moon and hence negligible for a system intended to operate on all clear nights (moon or not). While observations were not made fainter than $8^m0$, somewhat fainter observations could have been made on full moon nights and considerably fainter on nights with only city lights.

Besides being quite tolerant of background light, photometry is also tolerant of blockage of the sky in various directions by trees, etc. Although the best photometry is done near the zenith and photometry within 20 or 30° of the horizon is usually avoided, it is permissible to have one or more directions blocked entirely because where will be plenty of variable stars in the unblocked directions. Moreover, the sky does turn!

## 3.3   Physical Considerations

The observatory's main function, of course, is protecting the telescope and photometric equipment. When closed it should provide a barrier against rain, dust, snow, hail, birds, insects, and rodents. Also, when closed, the observatory should provide security from human entry, so provisions for locking the observatory securely should be included in the design. This also argues against a remote location for a small observatory, because it cannot be watched around the clock, as most big observatories are.

For convenient operation, the observatory should be designed so that it can be quickly set up for the night's operation and also quickly closed down. It helps if there is enough space to provide permanent storage for most things needed during observing: charts, equipment, tools, test equipment, and coffee.

A fair amount of light is needed for operation, to see the charts and equipment during photometry and to avoid collisions with the telescope parts. Thus, unlike most purely visual astronomy, where operating in the dark is important, many photometrists routinely observe with a surprising amount of light. When the equipment malfunctions, as it occasionally does, even brighter lights are useful for making the repairs (taking care to turn the high voltage to the photomultiplier down or off first). In colder climates, bad-weather maintenance, repairs, or improvements to the equipment are made more pleasant by a heater.

Having the electronic equipment on a movable cart or relay rack adds to the convenience in many installations. Some also have a roll-about table for the charts, etc.

## 3.4   Roof

A rotating dome with a slit is the traditional choice for a roof at most large observatories. While a dome has a few advantages, particularly in windy climates, it is not necessarily the best choice for the small observatory. Domes are very obtrusive and make security more difficult during absences. In astronomical photometry there is a lot of movement from one star to another all over the sky, so a roll-off or fold-off roof is more convenient in this respect. Domes tend to stick (especially when it is cold).

A roll-off roof tends to be cheaper and less likely to attract the curious, although it provides less protection from wind gusts. With a roll-off roof it is much easier to monitor sky conditions (e.g., approaching clouds). In areas prone to heavy snowfalls, a flat roof which does not shed snow should be avoided in favor of one with a steep pitch.

## 3.5   Roll-out Observatories

The roll-out or carry-out photoelectric observatory is becoming increasingly popular and has many advantages. The basic strategy is to store the telescope and photoelectric equipment in the garage, house, or classroom, and then move it out to a nearby, prepared observing site in a manner that allows quick set-up and equally speedy takedown. Properly approached, such an observatory can be set up just as quickly as a permanently housed observatory.

To allow for a quick set-up it is necessary that the outdoor observing site have an established pier or base such that the telescope, when placed on the base, does not require alignment. Similarly, the equipment, charts, etc. need to be organized on some sort of roll-out cart so that time is not lost in setting them up.

## 3.6   The Telescope

A telescope used for photoelectric photometry should be very solidly mounted, be well-aligned, have an accurate drive in right ascension, and be reasonably light tight.

The rigidity needed for photoelectric photometry is best provided by a very solid, permanent pier or pad, and a rigid, generously-proportioned mount. One must be able to flip the various controls on the photometer head without displacing the star from the center of a rather small diaphragm, nor must a slight wind be able to do this. During the centering of a star in the diaphragm to begin with, any vibrations introduced must dampen out quickly or else the task of centering becomes very difficult.

While the demands on the accuracy of the equatorial drive and alignment are not quite as severe as those required for long-exposure photography or spectroscopy, it is required that a star stay reasonably centered in the diaphragm for several minutes. As smaller diaphragms allow fainter stars to be observed (by reducing the sky background), accuracy approaching that required for long-exposure photography is required if one is to observe fainter stars. For home-built telescopes, a generously sized worm gear made specifically for telescopes by such manufacturers as Mathis, Byers, or Schmidt and driven by a synchronous AC motor is recommended. As a rule of thumb, the main gear should be almost as big as the diameter of the primary mirror or lens. A very accurate drive can be made quite cheaply by using the curved bolt tangent arm devised by Jones (1980). A drive corrector is very convenient, particularly as an aid to centering stars, although a few small installations do without this and rely on manual centering.

As it is convenient to operate with generous lighting on the observatory floor, the telescope needs to be well-shielded so that this light will not find its way into the photometer. Thus, "closed" telescope tubes are preferred.

## 3.7   Type of Telescope

If you already have a sturdy telescope, then this is the best type for you. If you do not, read on. Both refractors and reflectors have been successfully used for photometry. The refractor has the advantages of a closed tube, generally a conveniently placed eyepiece, and relative freedom from maintenance such as realuminization and realignment. The refractor also generally has a high $f$–ratio and long focal length, both helpful in photometry. On the minus side, the glass objective of the refractor absorbs much of the ultraviolet light and does not bring light of different colors to the same focus. If it is intended that most of the photometry will be in a single color such as V of the UBV system (and much very useful photometry is done in a single color), then these are not serious drawbacks. For a given aperture, refractors tend to be more expensive and their long lengths require larger observatories.

The Newtonian reflector has been the favorite choice of amateur astronomers for many years, as it is relatively inexpensive to build for a given aperture, not hypersensitive in alignment, and not overly sensitive to stray light. Many very successful photometric installations use a Newtonian telescope. Light of different colors comes to the same focus and there is no lens to absorb any ultraviolet light. As the eyepiece is at the top end, a long focal length and high $f$–ratio are difficult to achieve with a Newtonian and still have convenient (or even safe) access to the photometer head. Thus, with a Newtonian telescope of low to medium $f$–ratio, an especially compact photometer head is called for; or, at the expense of introducing a lens, a Barlow can be used to give a higher $f$–ratio and longer (effective) focal length.

A Cassegrain telescope combines the refractor's good features of a high $f$–ratio, long (effective) focal length, and convenient photometer position with the Newtonian's absence of lenses. A Cassegrain is more expensive to buy (or more difficult to build) than a Newtonian, however, and is bothered by stray light if it is not properly baffled. A Cassegrain is also more sensitive in alignment and its focus changes more with temperature. If one were starting from scratch and photometry were to be the main observing program, then a Cassegrain would be a good choice, although others would be acceptable. It is particularly convenient if the secondary mirror can be moved to achieve focus, thus allowing the photometer head to be solidly fastened to the tailstock.

Schmidt-Cassegrain and similar telescopes, such as those manufactured by Celestron, Mead, and others, are almost ideally suited to photoelectric photometry at smaller observatories. Their compact size makes them a favorite for roll-out or carry-out observatories, and even in permanent observatories a smaller, less expensive structure is possible. Their high $f$–ratios and convenient eyepiece/instrument location are well suited to photometry. The ability of most of these telescopes to have a fixed instrument load with the primary being moved for focus is a significant advantage, as it allows the photometer head to be solidly mounted on the telescope. While the corrector plate does absorb some ultraviolet light, ultraviolet photometry has so many difficulties at lower altitudes that this is not a serious loss. Moreover, the corrector plate allows a sealed telescope—an advantage of considerable practical significance. All things considered, if one were to buy a ready-made telescope for photoelectric photometry, then a telescope of this type is probably the best choice.

## 3.8 Aperture

A large aperture is not required for highly successful and useful photometry of variable stars. Many of the naked eye stars are variable and much useful photometry has been done with 6– and 8–inch telescopes. Experience has shown, somewhat surprisingly, that the amount of useful photometry done by amateurs tends to be inversely proportional to aperture size. In a study by Abt (1980), the cost-effectiveness of the optical telescopes at Kitt Peak National Observatory in terms of publications and citations was evaluated. He found that the smaller telescopes were by far the most productive.

In the choice of aperture, much depends on the intended observational program. Many Algol-type and RS CVn-type binaries, Mira-type variables, etc., are nicely handled by an 8–inch telescope. If RR Lyrae-type variables or dwarf novae are to be observed photoelectrically, then even a 12–inch telescope is not large enough. The beginner is strongly advised to start small and to note that some of the most successful amateurs of long standing and high productivity prefer an 8–inch to a larger telescope because it is easier to handle and maintain and gives them more than enough stars to observe. However, if the budget permits, telescopes of 10– to 14–inch aperture obviously will allow a greater variety of stars to be observed.

## 3.9 Mounts

All the various types of mounts have been used successfully for photometry if they were sufficiently solid in construction. Mounts which avoid the need for reversing the telescope when crossing the zenith are preferred for photometry, since much of the observing is done near the zenith. Two-pier mounts such as the yoke and the cross-axis are very solid, a prime consideration in photometry; although they sometimes block the pole, telescopes are difficult to handle in the polar region anyway, so the loss is not a serious one and plenty of sky remains. A fork mount, if it is very solid, is also a very good choice, except it tends to be especially poor in terms of instrument clearance when operating near the celestial pole, although, as pointed out above, this is not a great loss.

The importance of a solid mount and accurate drive in photometry cannot be emphasized too strongly. For smaller observatories, the smallest usable diaphragm is not usually determined by seeing conditions, as is the case with large observatory telescopes, but by the solidness of the mount, the accuracy of the drive, and the ability to center a star smoothly. At observing sites with relatively bright skies, the constraint on how faint a star can be observed is the contribution of the background light. Improvements in the mount and drive can allow a smaller diaphragm to be used and thus fainter stars to be observed. Such improvements can be cheaper and more effective than purchasing a larger telescope. A modest telescope on an immodest mount is perhaps the ideal! Thomas Mathis makes solid mounts for use with the popular Schmidt-Cassegrain telescopes. The "small" telescope mount made by DFM Engineering (Melsheimer and Genet 1984) is the last word in rigidity and is ideal for photometry except for its not insignificant cost.

## 3.10   Finder

The most important telescope accessory is a good finder.  In photometry, the
finder is used mainly to find the star and to center it very roughly, not for guiding
such as is done in photography.  The finder needs to have a low $f$–ratio, so that
it has a good field of view, and needs to have as large an aperture as one can
reasonably afford, so that rather faint field stars can be seen.  A finder used with
photoelectric photometry is particularly critical, receives much use, and generally
should have a lower $f$–ratio and larger aperture than one might otherwise think
necessary.  In fact, the criteria for a finder used with photoelectric photometry are
essentially the opposite of those for a photographic guider.  A 4– or 6–inch ($f/5$)
finder is not too large, while a 2–inch finder is really smaller than desirable.

It is especially convenient (even almost necessary) to have the finder's eyepiece
close to the eyepiece(s) on the photometer.  The finder needs to be mounted solidly
on the main telescope with convenient provisions for adjustment, preferably along
two perpendicular axes.  The normal three-screw ring mounts typically used on
finders leave much to be desired in these respects.  Perhaps an even better approach
is to permanently mount the finder on the main telescope and have an eyepiece
reticle adjustable in X–Y.

## 3.11   Setting Circles

Some observers find generously-sized, well-lighted setting circles to be an aid in
locating variable stars, while other observers never use or don't even have setting circles.  As most organized photoelectric observing programs require many
repeated observations of the same stars, one soon becomes very familiar with the
location of the variable and its comparison star; the use of setting circles or even
finder charts is not required except when a new star is added to the program.  It
might be noted, however, that, under conditions of a bright moon and slight haze,
setting circles do really help in finding even familiar stars and are a necessity
for finding unfamiliar stars.  For city observing sites, good setting circles are a
necessity at all times.

## 3.12   Fine Adjustments

Some observers use electrically operated fine adjustment of the telescope in right
ascension and declination, while other observers prefer mechanical adjustments.
If the clutches are smooth in operation and the telescope is well-balanced, it may
even be possible to center a star without fine motion controls.  While centering
in right ascension can be done by changing the drive speed and thus avoiding
problems with backlash, centering in declination involves frequent reversal in direction, so care must be taken to avoid problems with dead zones or backlash.  A
long tangent arm is a good solution to this problem.

Considering the frequency with which stars must be centered in making photoelectric observations, electrically-operated fine motion controls are a good investment indeed.

# Chapter 4

# THE PHOTOMETER HEAD

## 4.1  Introduction

For purposes of organization, the photometer head can be considered as that part of the photometric system which is attached to the telescope. Not included are the telescope itself and the supporting electronics. Included are the various diaphragms, mirrors, lenses, and filters in the optical path, and the device that converts the light into an electrical current, such as a photomultiplier or photodiode.

It should be realized that in some installations portions of the supporting electronics are physically mounted on or are an integral part of the photometer head. In most installations, however, the electronic equipment is entirely separate from the photometer head itself, so the distinction is still a useful one.

Commensurate with the central importance of the photometer head, this chapter is somewhat detailed. We begin with a brief overview of the entire photometer head and then follow with a discussion of each component. We conclude with relevant physical considerations, criteria for the selection of a photometer head, and specific examples of several photometer heads.

## 4.2  Overall Photometer Head

Before considering each component of the photometer head in detail, it would be useful to describe briefly all of the components which comprise a complete head. Such a head is shown schematically in Fig. 4–1. Individual designs may not contain all these components or may include them in a different fashion, but the diagram illustrates the main functions involved.

Starlight from the telescope objective enters the previewer section at the top of the head. A small mirror can be moved to deflect the light to a wide-angle eyepiece. This eyepiece, which often contains an illuminated reticle, is used to verify the correct star field and center the star more accurately than can be done with the finder but less accurately than with the microscope or postviewer, which comes later. If one has a large aperture finder, then a previewer would not be necessary unless one wanted to locate very faint stars.

Once the star is centered roughly, the mirror is moved out of the light path so that the starlight is focused in the plane of the diaphragm, which is used to exclude light from stars other than the one being observed.

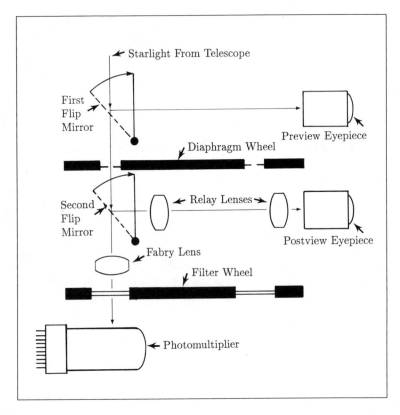

**Figure 4–1.** *Overview of photometer head shows previewer, diaphragm, postviewer, filter, Fabry lens, and photodetector. Most photometer heads have these essential ingredients, although some designs dispense with either the previewer or postviewer.*

A second mirror then can be moved into position to deflect the light to a microscope (postviewer) of moderate power which is focused on the plane of the diaphragm. This is useful for making small adjustments in the pointing of the telescope so that the star will be properly centered in the diaphragm. Once centered, this mirror is moved out of the light path.

The light then passes through the filter, which often consists of one or two pieces of colored glass. The filter allows light in one wavelength band to pass through, while rejecting light in other bands. Photometric measures with a sequence of different filters let the observer determine the relative brightness of a star at various wavelengths. With such information one can estimate stellar temperatures, interstellar reddening, etc.

The Fabry lens then concentrates the light onto the photosensitive surface in such a way that small movements of the star within the diaphragm do not affect the measurements.

A dark slide (not shown) is sometimes placed in the light path as it enters the photodetector housing. This keeps the detector in the dark, and hence protected, when the system is not in operation or when the detector in its housing is removed from the main head.

The photodetector converts light falling onto it into minute electrical pulses or a weak current. It can be one of two basic types: a photomultiplier or a photodiode. It is sometimes placed in a cold box to improve its performance.

## 4.3  Previewers

Some photometer heads include a provision for viewing the image from the main telescope before it reaches the diaphragm. This is usually done by using a flip mirror or insertable prism to deflect the light beam off to an eyepiece on one side.

In some head designs a previewer is the only provision for viewing the star. If this is the case, great care must be taken to insure that, when a star is centered on the reticle of the previewer's eyepiece and the previewer mirror is flipped out of the way, the star indeed will be accurately centered in the diaphragm. While this has been successfully done in a few designs, it is generally difficult to do in "homebrew" photometers and has been a fatal flaw in a number of designs. Thus, the inclusion of a microscope (postviewer) in homebuilt photometers is one of our strongest recommendations.

If a head design does incorporate a microscope for viewing the diaphragm and star from behind, then a previewer is not an absolute necessity. A number of photometric systems with large solidly mounted finders do not incorporate a previewer and experience no difficulties at all. Even with a good microscope (postviewer) and a large solid finder, however, there are several reasons for considering a previewer anyway. First, it allows the main telescope to be used for casual viewing, photography, etc., without removing the photometer head. Second, it lessens to some extent the necessity for changing from larger to smaller diaphragms to acquire a star initially. Third, it places less demand on the finder, although at the expense of a three-step (finder, previewer, microscope) centering process. Finally, it allows one to see much fainter stars in a reasonable field-of-view.

## 4.4  Diaphragms

The diaphragm, which is placed in the focal plane of the telescope, is used to exclude light from stars other than the one being measured. It is also used to minimize background light from around the star being measured. Stars as viewed in telescopes are not points of light; they appear as finite areas of light due to the diffraction produced as light enters through a finite aperture (the telescope objective). Furthermore, changes in atmospheric refraction cause the star image to have a finite size (the seeing disc) and also to jump around somewhat. On large telescopes, the atmospheric effects rather than the diffraction effects usually determine the lower limit on how small the diaphragm can be. For smaller, less massive telescopes, the accuracy of the drive in right ascension, the alignment of the polar axis, and the effects of wind can easily become the limiting factors in the smallness of a diaphragm. Also to be considered is the difficulty of initially positioning a star in the diaphragm, the task being more difficult with small diaphragms.

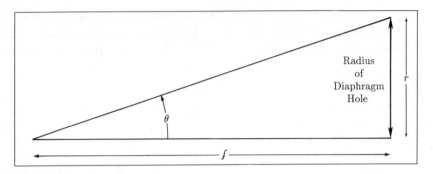

**Figure 4-2.** *The relationship between physical diaphragm size and angular coverage in the sky can be determined geometrically if the focal length of the telescope is known.*

Diaphragms are usually characterized either by the linear diameter of the opening or by the angular diameter of the sky one can view through that opening. These two parameters are equivalent for any particular telescope, and their relationship depends only on the effective focal length. This relationship can be derived from a simple geometric argument. Consider the situation shown in Fig. 4-2. For small angles, as is the case here, the approximation that the tangent of the angle is equal to the angle in radians is a very good one. Thus, the angle $\theta$ (in radians) is equal to the diaphragm radius $r$ divided by the focal length $f$, where both $r$ and $f$ must be in the same units, such as inches or millimeters. This same relationship relates the diaphragm diameter, $d$, to the angle of sky, $\phi$, seen through the diaphragm. Thus,

$$\phi(\text{radians}) = \frac{d}{f}. \tag{4.4.1}$$

Since arcseconds are more customary angular units in astronomy than radians, it may be helpful to note that there are 206,265 arcseconds per radian. Thus,

$$\phi(\text{arcsec}) = 206,265\frac{d}{f}, \tag{4.4.2}$$

where again $d$ and $f$ both must be in the same units. If we had a telescope with a focal length of 40 inches and we wanted a diaphragm to cover 50 arcseconds of sky, the diameter of the diaphragm would be 0.0097 inches.

The foregoing discussion of the diaphragm illustrates well the interactions between the various parts of the total system. The minimum useful diaphragm size (angular coverage) is affected by the solidity of the mount and the accuracy of its alignment, the accuracy of the drive, the protection against wind provided by the observatory, seeing conditions, accuracy of finder alignment, etc. The physical size of the diaphragm for a given angular coverage is determined by the effective focal length of the telescope.

Successful photometry has been done with a single fixed diaphragm size, and the simplicity of this approach has much to recommend it. For telescopes with very solid mounts and accurate drives, however, it is convenient to have a number of different diaphragm sizes available to suit different conditions of seeing, close visual companions, moonlight, and star brightness. For telescopes with less solid mounts and less accurate drives, the diaphragm size will be determined by these characteristics as opposed to seeing conditions; for such systems a fixed diaphragm size is not unreasonable.

Most photometer heads, however, have some provision for adjusting the diaphragm size. There are two basic approaches. One is to have a single diaphragm, the size of which can be adjusted. The other is to move diaphragms of different sizes into position. The different diaphragms can be mounted on a slide, a wheel, or a drum. In these schemes, some provision is usually made for locking each diaphragm into place with some sort of detent or other mechanism so that its center will coincide with the center of the others. If one uses a multiple diaphragm approach, it is useful to include one that is extremely large for initial centering and adjustments, and one that is completely closed for shutting out light when the system is not in use.

In making photometric observations, one usually moves the telescope off slightly for the sky background reading. Alternatively, the offset to the sky can be made from the star being observed by moving the diaphragm instead. This can be done manually or automatically. While it is generally more efficient than moving the entire telescope, it may move the light beam to a different part of the photocathode and thus introduces a systematic error into the measurements. Vignetting of the light beam could occur at the offset position, thereby producing a similar systematic error. One needs to check that measurements made with the diaphragm in both the regular and offset positions truly give the same results!

On bright moonlit nights there is no problem centering a star in the diaphragm of telescopes equipped with a microscope (postviewer) because the sky is sufficiently bright to show up as a disk in the eyepiece. On clear, dark nights, however, it is often difficult to see the edges of the diaphragm and thus know where to center the star. There are a number of remedies to this problem. One approach is to place an illuminated reticle in one of the diaphragms (a large one), center the star on this, and then switch to one of the smaller diaphragms. This requires accurate alignment of the diaphragms and a very positive detent mechanism. Another approach is to backlight the diaphragm. This works quite well, but care must be taken to be certain that the light does not get to the photomultiplier itself. Backlighting can be combined very effectively with an illuminated reticle in the microscope eyepiece. In this approach the backlight is used just to check the centering between the diaphragm and the illuminated reticle in the microscope eyepiece. Then, as long as the diaphragm is not changed, the backlight is no longer required and the star can be centered just on the microscope eyepiece reticle, with knowledge that this will assure centering in the diaphragm also. Besides being rather convenient, this approach reduces the danger of stray light entering the photomultiplier. Finally, one can introduce light into the main telescope itself—an approach that is simplicity itself and used by many.

The diaphragm must not be backlit or have light introduced into the telescope while a measurement is being made. Some systems use a microswitch, in conjunction with the microscope flip mirror mechanism, to turn off the backlight automatically when the mirror is not in the optical path. A number of the telescopes at Kitt Peak National Observatory which introduce light into the telescope have a prominent red light on the tailstock to remind the astronomer to turn the light off before making measurements.

Diaphragms for high $f$–ratio telescopes can be made by drilling a hole in a thin piece of metal. Very thin brass sheet works quite nicely for this. The smallest drill—a #80—gives a hole 0.0135 inches in diameter. For the smallest holes on a short $f$–ratio telescope this may not be small enough. In this case a much larger hole can be drilled in the brass sheet and a needle or pin used to make a hole in a piece of good quality foil; this piece of foil then can be cut out and glued over the larger hole in the brass sheet.

## 4.5   Microscopes (Postviewers)

While not an absolute requirement, the incorporation of a postviewing microscope into the simple photometer is one of the most worthwhile additions because it allows positive assurance that the star to be measured really is centered in the diaphragm and that other stars are not included. For large fixed-size diaphragms it allows nearby stars to be excluded by moving the target star somewhat off-center, although this is not particularly recommended. A microscope also allows one to ascertain positively that the diaphragm is in the focal plane of the telescope. We cannot recommend a postviewing microscope too strongly.

The optics and focusing of the microscope have nothing to do with the $f$–ratio or focal length of the main telescope; the microscope is actually an independent optical system. A flip mirror or prism often is used to view the diaphragm with the microscope. Some microscopes use a single lens plus an eyepiece to view the diaphragm, while others use two relay lenses plus an eyepiece. The relay lenses move the focal plane to a second, more accessible location. In order for relay lenses to operate properly, it is necessary that the first relay lens be at the proper distance from the diaphragm required to produce a parallel bundle of light. This can be achieved by removing the second relay lens and eyepiece, looking through the eyepiece hole with a small telescope (such as a finder) already focused on infinity, and moving the first relay lens until the diaphragm can be seen clearly in focus. The eyepiece is then replaced and set mid-range, and the second relay lens is inserted and adjusted for best focus. Overall focus is then achieved by adjusting the secondary mirror (in Cassegrain systems which have adjustable secondaries), by moving the entire photometer head relative to the telescope, or by moving the diaphragm and microscope relative to the telescope.

## 4.6   Fabry Lens [1]

Because of the effects of atmospheric seeing, improper centering, and irregular drive, a star image will seldom stay fixed in the center of the diaphragm for

---

[1] Written by John P. Oliver

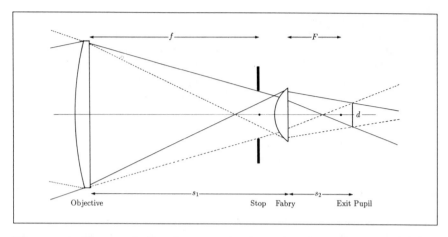

**Figure 4–3.** Diagram showing the geometric relationships for a Fabry lens. Note that the diaphragm does not have to be at the focus of the Fabry lens, and placing it there often needlessly wastes space.

any appreciable time. Rather, it will tend to wander in the focal plane. Since most photometric detectors have some degree of nonuniformity on their active surface, this wandering gives rise to a variable output signal. The Fabry lens is used to relieve this problem.

In an astronomical telescope the aperture stop is the objective lens or mirror. The objective, therefore, also defines the entrance pupil of the optical system. In order to reduce the effects of background sky brightness, the field of view of an astronomical photometer usually is limited by a diaphragm placed in the focal plane of the objective. If a lens is placed after the diaphragm in the optical train, the exit pupil of the system will be the image of the objective formed by this lens. Since every ray passing through the entrance pupil must pass through its conjugate point in the exit pupil, a detector placed at the exit pupil will see an immobile beam, unaffected by image motion in the focal plane of the objective. Such a lens is usually called a Fabry lens. The geometrical relationships among these various components are shown in Fig. 4–3 and are given by the formulas

$$\frac{1}{F} = \frac{1}{s_1} + \frac{1}{s_2} \tag{4.6.1}$$

and

$$\frac{b}{D} = \frac{s_2}{s_1}, \tag{4.6.2}$$

where $D$ is the diameter of the objective, $F$ is the focal length of the Fabry lens, $b$ is the diameter of the spot on the surface of the detector (the exit pupil), $s_1$ is the distance between the objective and the Fabry lens, and $s_2$ is the distance between the Fabry lens and the detector. Now, if the Fabry lens is placed close to the diaphragm and $\frac{b}{D} \gg 1$, it follows that $s_1 \cong f$ and $s_2 \cong F$, where $f$ is the focal length by the objective. Therefore, we have

$$F = b\frac{f}{D} \tag{4.6.3}$$

as a useful approximation.

It should be noted that no restriction has been placed on the exact location of the Fabry lens *vis-à-vis* the telescope objective. Frequently, drawings of photometers show this lens placed with the diaphragm at one focus of the Fabry lens. This is not only unnecessary but also frequently wasteful of space.

Simple plano-convex lenses from Edmund Scientific make acceptable Fabry lenses unless ultraviolet photometry is contemplated, in which case a quartz lens might be considered.

The proper functioning of the Fabry lens can be checked by moving a star about in a large diameter diaphragm and examining the constancy of the output.

## 4.7   Dark Slides

As will be explained in the next section, many photodetectors can be damaged or even ruined completely if the photosensitive surface is exposed to too much light when the high voltage is turned on. Potential danger exists, therefore, when the sky is bright in the early evening or approaching dawn, when the observer is searching for a variable star near the full moon, or whenever full lights are turned on in the observatory. To prevent such damage there should be some provision for keeping the photodetector in darkness. If provisions have been made for changing the diaphragm and/or filters, then this can be accomplished to some extent by including a blank (opaque) position to cut off the light. The previewer mirror also can serve this function. Another approach is to include a slide or shutter specifically for this purpose. This is called a dark slide. Even with no high voltage applied to the photodetector, extremely intense light, such as direct sunlight, can degrade many photodetectors. Therefore, when the telescope is not in use or is being used for nonphotometric purposes, it is convenient to have a dark slide on the photomultiplier housing itself. While convenient, many systems do not have dark slides.

## 4.8   Photomultiplier Tubes [2]

### 4.8.1   Introduction

All stellar photometers incorporate a sensitive detector in order to provide a measurable signal. This detector is usually a photomultiplier tube (PMT). A photomultiplier tube consists of a light-sensitive photocathode which emits photoelectrons at a rate modulated by the signal, and a multiplier section which amplifies the signal to provide a measurable current output. Photomultiplier tubes are sensitive to radiant energy from the ultraviolet (105 nm) to the near infrared (1100 nm). Since their inception in the late 1930's, no other device has been developed which can provide the amplification versus noise versus cost that is characteristic of a PMT. Their ability to detect single photons makes them particularly useful in low-light level astronomical applications.

The purpose of this section is to investigate the aspects of PMT design and performance as they pertain to stellar photometry. Since it is the photomultiplier

---

[2]Written by Robert C. Wolpert

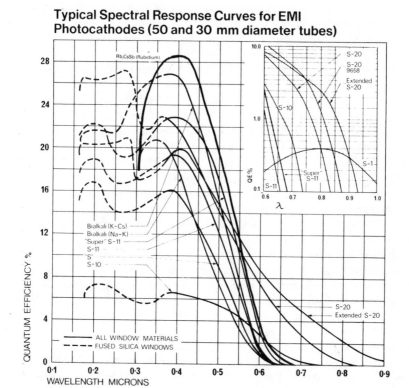

**Figure 4–4.** *Typical spectral response curves in terms of absolute sensitivity.*

tube which is converting low-level starlight into a useful signal, it should be chosen to provide optimum performance.

### 4.8.2 Spectral response

The two principle factors which determine the spectral sensitivity of a PMT are the type of photocathode material and the transmission characteristics of the front window.

### Photocathodes

There are two basic classes of photocathodes—opaque and semitransparent. The opaque photocathodes are commonly found in side-window PMT's such as the EMI 9781 and RCA 1P21. The photoemissive material is deposited on a nickel substrate which is located approximately 12 mm inside the glass envelope. Photoelectrons are emitted from the side of the photocathode that is struck by incident photons.

End-windows PMT's are usually designed with a semitransparent cathode. The cathode is deposited onto the inside surface of the window to a depth of 200 to 300 Å. Since the photoelectron is emitted from the side of the cathode opposite to the incident light, the thickness of the cathode becomes critical. If it is too thick, the photoelectrons will not be able to escape from the emissive layer; if it is too thin, light can be transmitted without creating a photoelectron and the PMT will have poor quantum efficiency.

Quantum efficiency, Q.E., can be defined as the number of photoelectrons emitted by the photocathode per incident photon, and is almost always described as a percentage. The Q.E. is dependent upon the wavelength of the light source and, as shown in Fig. 4–4, usually peaks between 350 and 450 nm. Typical peak quantum efficiencies are 20% to 30%, i.e., for every 100 incident photons, 20 to 30 photoelectrons are created. In low-light level stellar photometry, it is important to have as high a quantum efficiency as is reasonably achievable.

The photocathode material will determine the sensitivity of the PMT to light in the 300 to 1100 nm range. The basic photocathode is generated by the deposit of cesium and antimony in the form of $Cs_3Sb$. This will yield the basic S–4, S–5, and S–11 spectral response for UBV measurements, with a red cutoff at 650 nm. If potassium is added to form $K_2CsSb$, we obtain what is commonly referred to as a bialkali photocathode. It has improved blue sensitivity and lower dark current than the $Cs_3Sb$ cathodes at the sacrifice of a small amount of red response. The cathode resistivity is very high, so one must be careful if high cathode currents are expected or if the tube is cooled below $-10°C$. In stellar photometric applications, this is seldom a problem.

The S–20 (also known as multi-alkali or tri-alkali response) is obtained by adding sodium to form $Na_2KSbCs$. This extends the red cutoff to 850 nn, and special processing techniques will provide sensitivity to 960 nm. These are the extended red or "ERMA" PMT's. The S–20 tubes should be used for UBVR and similar photometric systems in which the stellar red response is measured. Near-IR response for the UBVRI system can be obtained by using a silver-oxygen-cesium (S–1) photomultiplier. These tubes will provide the best sensitivity between 900 and 1000 nm, but at the expense of sensitivity in the shorter wavelengths and a large increase in dark noise. The S–1 PMT's perform best when operated in housings cooled by dry ice or liquid nitrogen. Another common type of photocathode is the III–V, negative electron affinity materials such as GaAs and InGaAs. They have the highest quantum efficiency and flattest spectral response in the 300 to 900 nm region and produce very low dark counts when cooled between $-30°$ and $-60°C$. They also have several disadvantages. The ambient temperature dark counts are so high that the tubes must be used in cooled housings; they usually have maximum anode currents of only 50 nA, thus severely limiting the dynamic range of the photometer; their photocathode area is very small (even in tubes 2 inches in diameter) and therefore require accurate positioning and guiding; their life expectancy is likely to be shorter than with other photomultiplier types; and their cost is high. These limitations will generally preclude their use in amateur photoelectric photometers. However, where high sensitivity and low noise across UBVRI are needed, a well-cooled GaAs PMT cannot be beat. Hamamatsu makes a GaAs PMT with a photocathode of reasonable size and excellent efficiency over a very wide band width.

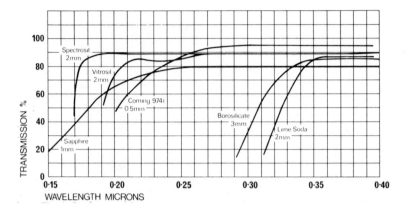

**Figure 4–5.** *UV transmission of typical photomultiplier windows.*

## Windows

Starlight (or any other light) must first be transmitted through the glass window or envelope of the photomultiplier tube before it can impinge upon the photocathode. It is the transmission of the glass which determines the ultraviolet cutoff characteristics of a photomultiplier tube. Fig. 4–5 shows typical window material transmission vs. wavelength. If it is desired to detect radiant energy below 300 nm, it is important to remember that all surfaces in the optical path must be capable of UV transmission.

### 4.8.3 Dynodes, Gain, and Rise Time

The multiplier section of a photomultiplier tube consists of a number of dynodes and the anode. The current amplification is achieved through the secondary emission characteristics of the dynode material. The secondary emission, $\delta$, occurs when an incident electron strikes the dynode with a kinetic energy sufficient to excite several electrons on the surface of the dynode material to an energy which will allow them to escape the potential barrier. We may therefore define $\delta$ as the number of secondary electrons emitted per incident primary electron. Since the energy of the primary electron is dependent upon the interdynode potential, we may change the amount of secondary emission by adjusting the voltage across successive dynode stages. There exists a primary electron energy for which the secondary emission is optimized.

The two most common dynode materials are CsSb and cesiated BeOCu. Typically, $\delta$ is 4 to 6 for CsSb dynodes and 3 to 4 for BeOCu dynodes. All else being equal, a PMT with CsSb dynodes will have a higher gain characteristic than a BeOCu photomultiplier. Most PMT's used in stellar photometers incorporate CsSb dynodes. Another type of secondary emitter is GaP:Cs. It has a

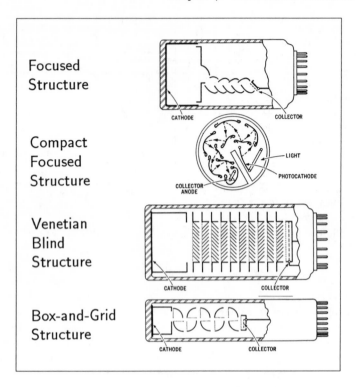

**Figure 4–6.** *Dynode chain configurations.*

$\delta$ of 15 to 25, but is much more costly. This amount of secondary emission is generally not required in stellar photometric applications. We can now define the amplification or gain, $G$, of a PMT as

$$G = \delta^N, \tag{4.8.1}$$

where $N$ is the number of dynode stages. (This represents the ideal situation in which all electrons created at the cathode are collected at the first dynode, and all secondary electrons are collected by the following dynodes in the multiplier section. The actual collection between cathode and first dynode is typically 90%. The collection efficiency between successive dynode stages is close to 100%.) Since we can alter the secondary emission of each dynode by adjusting the voltage applied between them, we can vary the PMT gain by altering the total high voltage applied between the cathode and anode. Photomultiplier gains are typically between $10^4$ and $10^8$. A 10–stage tube with $\delta = 4$ and 100 volts across each stage will yield a gain of about $10^6$, which is a good operating amplification for stellar photometry.

Pulse-counting (also known as photon-counting) techniques are very useful in low light level photoelectric photometry applications. This method is often recommended because:

1. The signal-to-noise characteristics are improved,

2. It is insensitive to photomultiplier thermal gain changes and D.C. leakage currents,

3. The calculation of the error in the signal and background is simple and exact, and

4. The necessary electronics are inexpensive and easy to build.

The only significant disadvantage is that, at high count rates, the readings must be corrected for error due to dead time, which is caused by the width of the pulse at the anode and the finite pulse pair resolution of the preamp and counter. As the mean pulse rate increases, a greater percentage of the output pulses will fail to be resolved. In a pulse counting mode with a PMT gain of $10^6$ and a band pass of $5 \times 10^{-9}$ seconds (FWHM), a single photoelectron will provide the following output pulse:

$$\frac{(1.6 \times 10^{-19}\text{coul.})(10^6)}{5 \times 10^{-9}\text{sec}} = 3.2 \times 10^{-5}\text{amps.} \tag{4.8.2}$$

If this current is run through a 50 ohm load, we will obtain a 1.6 mV signal. A preamp gain of 10 would allow a convenient discriminator setting of 16 mV. Most discriminators will have a pulse pair resolution of 30 nanoseconds or longer. The rise-time characteristics of a PMT will become significant only when it limits the pulse resolution of the following electronics. The rise time of a PMT is determined primarily by the type of dynode structure used in the tube. The four basic structures are shown in Fig. 4–6. The linearly and circularly focused (squirrel-cage) structures have rise times of only 1.5 to 3.0 nanoseconds (2.5 to 5 nanoseconds F.W.H.M.), depending upon the number of stages and the voltage applied. Venetian Blind and Box Grid structures will have typical rise times of 8 to 15 nanoseconds (20 to 45 nanoseconds F.W.H.M.), again depending upon the number of dynodes and the voltage. Obviously, the focused structures are preferred for fast pulse-counting systems.

### 4.8.4  Noise

The two basic forms of photomultiplier noise are the noise which is present in the signal and that which is present when the photomultiplier tube is shielded from all external light, i.e., the dark noise.

The signal noise, called shot noise, may be described more accurately as the random fluctuations of the signal. It is defined by the equation

$$i_k = \sqrt{2eI_k\Delta f}, \tag{4.8.3}$$

where $i_k$ is the r.m.s. shot noise current, $I_k$ is the total mean cathode current, $e$ is the electron charge ($1.6 \times 10^{-19}$ coulomb), and $\Delta$ is the bandwidth. The r.m.s. noise at the anode then should be

$$i_a = Gi_k = G\sqrt{2eI_k\Delta f}, \tag{4.8.4}$$

where $G$ is the PMT gain. (There is also a small contribution to the shot noise which is basically dependent upon the PMT collection efficiency and the first dynode gain. This noise contribution will be at a minimum when the collection between cathode and first dynode and the first dynode gain are optimized. It is quite small and is usually neglected.)

Equation (4.8.4) can be written as

$$i_a = \sqrt{2eGI_a\Delta f}. \tag{4.8.5}$$

Let us assume that the light of a 1st magnitude star collected by a mirror 6 inches in diameter is $1.12 \times 10^{-8}$ lumens at the cathode of a photomultiplier which has a gain of $10^6$, a cathode sensitivity of 50 microamps/lumen, and a dark current of 1.0 nA. The anode signal current from this star would be

$$I_a = (1.12 \times 10^{-8} Lm)(50 \times 10^{-6} ALm^{-1})10^6 = 5.6 \times 10^{-7} A. \tag{4.8.6}$$

Using equation (4.8.5), we find the r.m.s. noise at the anode in the signal would be

$$i_a = 2(1.6 \times 10^{-19} coul)(10^6)(5.6 \times 10^{-7} A)1sec^{-1} = 4.23 \times 10^{-10} A \tag{4.8.7}$$

for a 1 Hz bandwidth. Using the same equation but substituting the anode dark current for $I_a$, we obtain the r.m.s. noise in the dark current,

$$i_{d.c.} = 1.79 \times 10^{-11} A. \tag{4.8.8}$$

These results show that the ratio between the signal and the noise in the signal is theoretically

$$\frac{S}{N} = \frac{5.7 \times 10^{-7} A}{\sqrt{(4.23 \times 10^{-10} A)^2 + (1.79 \times 10^{-11} A)^2}} = 1346. \tag{4.8.9}$$

Since the dark noise contribution is significantly smaller than the noise in the signal for a 1st magnitude star, any improvement in the signal-to-noise ratio can be attained only by selecting a tube with higher quantum efficiency. Selecting a tube with twice the sensitivity (100 $\mu$A/Lumen instead of 50 $\mu$A/Lumen) will improve the signal to noise ratio by $\sqrt{2}$.

The dark noise of a PMT is composed of several components: thermionic emission, Cerenkov radiation from cosmic rays, radioactive materials in the PMT components, and ionization noise (after pulsing). Since ohmic leakage is DC, it usually can be ignored unless it is so great that it cannot be nulled; however, this is generally not a problem in contemporary photomultiplier tubes.

The largest contributor to PMT dark noise is the thermionic emission. Fortunately, there are several ways of significantly reducing this noise. The thermionic emission of electrons from a photoemissive surface is described by the Richardson equation:

$$N = AT^2 e^{-\phi/kT} cm^{-2} sec^{-1}, \tag{4.8.10}$$

where $T$ is the temperature in $K$, $\phi$ is the work function in electron volts, and $k$ is the Boltzmann constant. The $A$ term is defined as

$$A = \frac{4\pi emk^2}{h^3} sec^{-1} cm^{-2} K^{-2}, \tag{4.8.11}$$

where $e$ is the electron charge, $m$ is the electron mass, and $h$ is the Planck constant. Simple arithmetic gives a value for $A$ of $7.5 \times 10^{20}$ electrons $sec^{-1}cm^{-2}K^{-2}$. Note that this figure is actually slightly less for semiconductors, and the temperature dependence becomes $T^{5/4}$.

This clearly demonstrates that the thermionic dark noise is directly dependent upon the temperature of the photocathode and the photocathode area. Thus, we may expect that at room temperature an S–20 (tri-alkali) photocathode will emit 400 electrons $sec^{-1}cm^{-2}$, an S–11 photocathode will emit 100 electrons $sec^{-1}cm^{-2}$, and a bialkali photocathode will emit 30 electrons $sec^{-1}cm^{-2}$. This thermionic noise can be reduced significantly either by lowering the temperature of the photocathode, by reducing the photocathode area, or by using a defocusing magnet.

The magnetic defocusing lens is generally used in combination with a 50 mm (2 inch) diameter PMT in situations where a spot on the cathode less than 10 mm in diameter is actually being illuminated. The typical magnet is 50 mm in diameter and 6 mm thick, with a central circular aperture 10 mm in diameter. When placed adjacent to the front window of the PMT, it will defocus electrons emitted from the unused periphery of the photocathode, thus preventing them from striking the first dynode and contributing to the total noise. Such magnetic defocusing systems can decrease the ambient dark current to less than 10% of the unfocused dark current.

Cooled PMT housings are commercially available for all side-window tubes and end-window tubes up to 50 mm (2 inches) in diameter. They are desirable not only because they greatly reduce thermionic dark noise, but also because they provide temperature (and therefore gain) stability. Various models are designed to provide operating temperatures between 0°C and −100°C, but it is important to remember that colder is not always better. After reaching the optimum temperature for a particular photocathode, further cooling will not significantly reduce thermionic emission. It may, however, reduce the PMT gain, change the spectral sensitivity characteristics, and cause unnecessary stress on the base of the tube, often resulting in a glass crack near the pins. The S–1 photocathode is the only type which benefits from dry ice and liquid nitrogen temperatures of −78°C and −196°C, respectively. Most GaAs tubes are adequately cooled at −30°C, although the dark count will continue to decrease by cooling down to −60°C. The S–20 and S–11 photocathodes will reach optimum performance between −30°C and −40°C, which is typically provided by forced air-cooled and water-cooled thermoelectric housings. The bialkali photomultipliers are normally cooled between 0°C and −10°C, as further cooling will make the cathode more resistive and offer little dark noise improvement.

### 4.8.5   Cathode sensitivity and uniformity

Photomultiplier tube manufacturers commonly describe cathode sensitivity in units of microamps per lumen. This provides an easily measured system of classifying or grading the photomultipliers. A tungsten filament lamp operated at a color temperature of 2856°K is used as a light source, while the output is measured from the first dynode. Typical values range from 20 to over 300 microamps/lumen. While this reading is convenient for measuring the visible performance of a PMT, it is also necessary in stellar photometric applications to have an indication of the tube's performance in the blue, red, and near-IR spectral regions. This information is provided in the form of filter measurements.

Blue response is measured with a half-stock Corning CS–5–58 filter and is referred to as the Corning Blue (C.B.) number. This number will furnish a good indication of the sensitivity of the PMT at 420 nm. It is possible to estimate the quantum efficiency of bialkali photocathodes by multiplying the C.B. number by 2.5; therefore, a bialkali PMT is likely to have 25% Q.E. at 420 nm if the C.B. number is 10.0. Typical values range from 5.0 to 14.0.

Red response is measured with a Corning CS–2–62 filter, which peaks between 650 and 750 nm. Corning Red (C.R.) numbers are supplied with S–11, S–20, and Rb-bialkali PMT's to grade their relative red response. Wratten 86 and 88 filters are used for the measurement of near-IR sensitivities of S–20, S–1, and side-window tubes. The transmission of these filters peaks between 800 and 900 nm. For UBV measurements, the Corning Blue filter reading should be considered when selecting a PMT. Likewise, for UBVR work the C.B. and C.R. filter values should be considered.

Cathode uniformity describes the ability of the PMT to provide the same output current for equal light impinging on different areas of the photocathode. A non-uniform cathode could result in some troublesome magnitude variations if a focused star image is allowed to drift across the cathode area during measurement. For this reason photometers are designed to let the light beam cover a large portion of the effective photocathode diameter, thus averaging any non-uniformity.

Non-uniformity is caused by variances in collection efficiency, window design, and deposition of the photoemissive layer. Side-window tubes are usually the poorest for cathode uniformity and collection efficiency characteristics, due to the non-uniform electrostatic field between the cathode and first dynode. End-window PMT's 2 inches in diameter are generally regarded as the best in these parameters. Cathode uniformity (and usually sensitivity) can be improved by sandblasting the front window, thereby providing a diffusing layer.

### 4.8.6   Voltage divider design

A voltage divider network is necessary to provide the correct potential distribution between each element in the multiplier section of the photomultiplier tube. The current in the divider chain should be at least 10 times the mean anode current in order to ensure constant dynode voltages. In order to achieve the best possible signal-to-noise ratio, the voltage divider should be designed to maintain an adequate electric field between the cathode (K) and first dynode ($D_1$), thus optimizing the $K - D_1$ collection efficiency, and to provide a large first dynode gain. It is therefore recommended that the $K - D_1$ resistor value be twice the dynode chain standard resistor value for all PMT's 2 inches in diameter and smaller, or that a Zener diode be used to maintain the $K - D_1$ potential when the photomultiplier is operated at lower than normal voltages. Since stellar photometric applications rarely generate large mean anode currents or linearity problems, the voltage divider design is straightforward. Negative polarity high voltage is usually preferred, since it eliminates the possibility of power supply ripple triggering the discriminator in photon-counting systems. Also, it allows you to integrate the signal to obtain a DC analog output.

Voltage divider networks may be potted with RTV–11 (a silicone compound) to protect their components from moisture in humid environments. It should be noted, however, that this also makes any repair or modification of the divider chain much more difficult. Alternatively, a Humiseal coating may offer effective protection and still allow easy access to the components.

The power supply should be selected to provide well-regulated voltage with enough current to drive the dynode chain easily. The voltage requirement would depend upon the particular type of PMT used; but a 1500 volt, 1 mA power supply will suffice for most photometers. Obviously, the voltage polarity must conform to the design.

### 4.8.7 PMT selection

The manufacture of a photomultiplier tube is still considered an art. Because of the many variables involved in their design and manufacture, individual tubes will show a wide spread in parameters. Parameters quoted in catalogs are values to be expected with a "typical" tube. Because many are produced with superior characteristics, it is to the advantage of anyone anticipating the purchase of a PMT for stellar photometry to have the manufacturer select the tube to certain specifications.

Most professional astronomers will have the PMT selected for high quantum efficiency at the desired wavelength and for low dark current and/or dark count. A selected tube may have as low a dark count at ambient temperature as some unselected tubes have when cooled, and it is much less expensive to have a tube selected than to purchase a cooled housing. Some selected tubes provide only a few counts per second background when operated inside a cooled housing.

Multichannel photometer systems may require similar PMT's to be selected to a particular gain spread. This can be accomplished by specifying a particular voltage range required to provide a specified gain or output signal.

It is also possible to match tubes in quantum efficiency at specified wavelengths. This is usually accomplished through a monochromator calibration in which the selected tube is compared to the response of a traceable standard at different wavelengths, or through the matching of filter readings. The monochromator readings are the most reliable, but even these measurements are accurate only to within about 10%.

Some PMT manufacturers will also furnish a count rate plateau curve which plots the background and/or signal counts against the applied voltage. For photon-counting systems it is recommended that a tube be operated within the voltage range of the plateau so that the change in background signal will be small for small voltage changes. Plateau curves can be determined at cooled or at ambient temperatures.

### 4.8.8 Operating precautions

A photomultiplier tube will provide many years of consistent reliable service when operated correctly. There are several precautions every PMT user should be aware of:

1. Never expose the PMT to ambient light while high voltage is applied. This may cause irreparable damage to the tube: usually in the form of excessive dark current, low gain, or reduced sensitivity.

2. Avoid placing a PMT in a helium environment. Helium will diffuse through the glass envelope and will be detrimental to the noise and after-pulsing characteristics.

3. Never leave the photometer head attached to a telescope exposed to direct sunlight. The heat and intense light will "burn" the photocathode even without applied high voltage, resulting in loss of sensitivity.

4. Always observe caution when inserting a PMT into a socket. The pin base is the most susceptible place for a glass crack. Cooling and warming the PMT should be done gradually to avoid unnecessary stress on the glass base of the tube.

5. Never lay a photomultiplier tube on a table. They always find some way of rolling off; and the better the tubes, the more prone they are to rolling.

I would like to thank Louis A. Evangelist of EMI Gencom Inc. for his many hours of patient instruction and his insight into the performance of photomultipliers in stellar photometers. He has developed a stellar photometer suitable for use at amateur and college observatories.

## 4.9  Detection of Faint Stars with Photomultipliers [3]

We will now derive the equations which allow the prediction of the faintest detectable star, and the limiting magnitude given various precision requirements, for any photometric system. This treatment has been adapted from Baum (1962), Pecker (1970), and Whitford (1962).

### 4.9.1  Photon Statistics

If, in our photometer, we possessed a perfect detector that would record the arrival of each incident photon, then our measurement of stellar flux would consist simply of waiting for a sufficient number of quanta to arrive to give us any desired accuracy. In practice, the detector responds to a fraction of these, $q$, called the quantum efficiency. Since both the arrival and the detection of each photon occur randomly, to a close approximation, we can obtain $v$ photoevents in an interval $t$ if the average rate of photon arrival is $n$. This gives the relation

$$v = nqt. \qquad (4.9.1)$$

The standard deviation, $\Delta v$, will be

$$\Delta v = \sqrt{v} \qquad (4.9.2)$$

---

[3]Written by Richard Berry

and the signal to noise ratio, $\frac{S}{N}$, will be

$$\frac{S}{N} = \frac{v}{\Delta v} = \sqrt{v} = \sqrt{nqt}. \tag{4.9.3}$$

The background sky contributes $n_b$ photons to the $n_s$ photons from the star and, in order to subtract the background, one-half of the observing time must be spent measuring the sky radiation. The detector also contributes $n_d$ false events, usually called dark counts or dark current. Thus we have

$$\frac{S}{N} = \frac{0.5qn_s t}{\sqrt{(n_d + 0.5qn_s + qn_b)t}}. \tag{4.9.4}$$

If the background and dark current are small, this relation will reduce to the original simple relationship. However, if the background is the strongest source, we have the relations

$$n_b \gg n_s \gg \frac{n_d}{q}. \tag{4.9.5}$$

In this case, it follows that

$$\frac{S}{N} = \frac{\sqrt{q}n_s\sqrt{t}}{2\sqrt{n_b}}. \tag{4.9.6}$$

If the time of observation is held constant, note that

$$\frac{S}{N} \propto n_s \tag{4.9.7}$$

and

$$\frac{S}{N} \propto \frac{1}{\sqrt{n_b}}. \tag{4.9.8}$$

To attain a given signal to noise ratio,

$$\frac{S}{N} \propto \sqrt{t}, \tag{4.9.9}$$

indicating that, for faint objects, attaining higher precision requires a larger amount of time.

### 4.9.2 Additional parameters

Light from the star must enter the diaphragm. In order to ensure that we are getting all the starlight, the diaphragm must be several times larger than the size of the stellar image. The angular size of the diaphragm, $a$, in radians, should be as small as possible to minimize $n_b$.

For a rough calculation of the limiting magnitude of any system, we will define detectability such that

$$n_s qt \geq \frac{S}{N}\sqrt{a^2 n_b qt}; \tag{4.9.10}$$

when

$$\frac{S}{N} \sim 1. \tag{4.9.11}$$

Using an objective of diameter $D$, whose area collects both star and background photons

$$D^2 n_s q t \geq \frac{S}{N} \sqrt{a^2 D^2 n_b q t}; \tag{4.9.12}$$

and, including the dark current events, this becomes

$$D^2 n_s q t \geq \frac{S}{N} \sqrt{a^2 D^2 n_b q t + n_d t} \geq \frac{S}{N} a D \sqrt{(n_b q t)(1 + d)}, \tag{4.9.13}$$

where $d$ is the ratio of the dark current to the sky signal.

For a star to reach a given signal to noise ratio, the ratio of stellar intensity to background noise will be

$$\frac{n_s}{n_b} = \frac{S}{N} a \frac{\sqrt{1 + d}}{D \sqrt{n_b q t}}. \tag{4.9.14}$$

### 4.9.3  Limiting magnitudes

To compute limiting magnitudes, we can cast these equations into suitable form. First, convert magnitudes per steradian sky brightness into magnitudes in the diaphragm, $m_a$, with the relation

$$m_a = m - 2.5 \log a^2 = m - 5 \log a, \tag{4.9.15}$$

where $m$ is the sky brightness in magnitudes per steradian.

From the definition of magnitude, the number of photons $n$ can be related to the magnitude $m_o$ by

$$m_o = -2.5 \log n + \text{constant}. \tag{4.9.16}$$

Transforming equation (4.9.14) into log form and converting to magnitudes, the threshold $(S/N = 1)$ is

$$m_o - m_a = -2.5 \log a - 2.5 \log \frac{S}{N} + 1.25 \log t - 1.25 \log(1 + d) +$$
$$1.25 \log q + 1.25 \log n_b + 2.5 \log D. \tag{4.9.17}$$

Finally, putting the terms in logical order we have

$$
\begin{array}{lll}
m_o = 0.5 m_a & \text{Background, mag/ster.} & \\
\quad -2.5 \log a & \text{Diaphragm size in radians} & \\
\quad +1.25 \log t & \text{Integration time} & \\
\quad +1.25 \log q & \text{Quantum efficiency} & \\
\quad +2.5 \log D & \text{Telescope aperture in cm.} & (4.9.18) \\
\quad -1.25 \log(1 + d) & \text{Relative dark current} & \\
\quad -2.5 \log \frac{S}{N} & \text{Required } \frac{S}{N} \text{ out} & \\
\quad +\text{Constant} & \text{About 7.5 for blue light.} &
\end{array}
$$

Other factors such as the transmission of the optics and the passband of the filters are included in the constant, derived from systems in which all the parameters were known.

### 4.9.4 Dominant factors in limiting magnitude.

Referring to equation (4.9.18) and assuming that the same fractional improvement in each parameter is equally easy, we pick out $a$, $D$, and $\frac{S}{N}$ as being the most influential. To gain improvement proportional to the change we apply, we can

1. Reduce the diaphragm (i.e., a more stable mount or accurate drive),

2. Increase the aperture (very expensive), or

3. Reduce the $\frac{S}{N}$ requirement (i.e., plan the experiment more carefully).

To gain improvement proportional to the square root of the change, we can pick on $m$, $t$, $q$, or $d$ and

1. Reduce the background (i.e., find a darker site),

2. Increase integration time (i.e., decrease program size),

3. Increase $q$ (a major research effort), or

4. Find a better tube or refrigerate the one we have.

The implication is that, for optimum work near the limit (with any value of $t$, $q$, or $d$), the site must be improved. Barring this, we must

1. Increase integration time,

2. Decrease the $\frac{S}{N}$ requirement, or

3. Avoid working near the limit.

The last suggestion has the merits of practicality, economy, and simplicity.

### 4.10 Photodiode Detectors [4]

In his original observations of the light intensity of the moon in 1907, J. Stebbins used a solid-state selenium detector. While these detectors were soon replaced by the more sensitive and reliable vacuum photocells, and these in turn by the photomultiplier, recent advances in both solid-state sensors and in very low current amplifiers have resulted in the return of the solid-state sensor to astronomical photometry. It is interesting to note that H. L. Johnson, who did much of the pioneering work in the initial application of the photomultiplier to astronomical photometry, also did pioneering work in the application of photodiodes to this same field.

---

[4]Written by Gerald Persha

| TABLE 4-1 |
|---|
| COMPARISON BETWEEN PHOTOMULTIPLIERS AND PHOTODIODES |

| Characterstic | Photodiode/Amplifier | Photomultiplier |
|---|---|---|
| Stability | Very insensitive to strong light,temperature changes or mechanical shock | Variable with time, temperature, light, power supply fluctuations, and shock |
| Spectral response | 200 to 1100 nm in one detector | 200 to 1000 nm possible with use of several units |
| Response time | Slow when configured for greatest sensitivity | Fast in photon counting configuration |
| Sensitivity | Approximately $10^{-16}$ watts | Will count individual photons |
| Power supply | $\pm5$ to $\pm18$ volts | 600 to 3000 volts |
| Size | Can be housed in a package less than 1 cm$^3$ | With shielded housing normally $1\frac{1}{2}$" in diameter and 5" in length |

The photodiode photometer uses a silicon PIN photodiode in place of the more traditional photomultiplier tube as the light detecting device. Recent advances in the design and manufacture of photodiodes have improved their performance specifications to a level necessary for astronomical applications. While not equaling photomultiplier tube performance in low light sensitivity (photomultiplier tubes can detect individual photons, whereas this is inherently impossible for photodiodes), photodiodes do offer some distinct advantages, as shown in Table 4-1.

The PIN photodiode is very similar to the common PN diode used in countless day-to-day applications, with the major differences being that a thin wafer of very high-resistance silicon is used to make the PN junction and the junction is open to light. Fig. 4-7 shows the basic PIN photodiode configuration.

As seen from Fig. 4-7, the letters PIN in the name represent the three distinct regions which make up the photodiode. The P-type region is made by diffusing an electron-acceptor material into the high resistivity silicon wafers; this produces a large number of positively-charged free holes in the conduction band of this region. Likewise for the N-type region, an electron-donor material is diffused into the opposite side of the wafer; this produces a large number of negative free electrons in the conduction band of this region. The region between the P-type and N-type diffusion regions is called the I or Intrinsic layer.

Both sides of the silicon wafer are still electrically neutral since the charges in the conduction band are balanced by an equal number of oppositely charged atoms locked into the silicon crystal, the valence band. The charges in the two conduction bands attract each other, however, and some migrate to the other side in an attempt to fill the silicon crystal uniformly. When this occurs, electrical neutrality is destroyed and a small electric field is generated from the

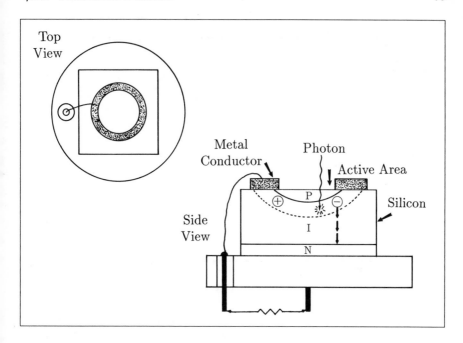

**Figure 4–7.** *The PIN photodiode. P– and N–type materials are separated by an insulating (I) layer.*

N to the P layer which eventually inhibits further migration. One might think at this point that a measurable potential could be found across the junction, but this is not the case.

The electric field across the Intrinsic layer depletes a region near the P–type layer of all free electrons or holes. In some devices, the depletion region may extend the entire width of the P–type layer. Light energy entering the junction and being absorbed in the depletion region creates electron-hole pairs which are separated and pulled to the opposite sides of the junction along the electric field lines. If the two sides of the junction are connected by a wire, a current will flow through it which is directly related to the rate of electron-hole pair creations in the I layer. Fig. 4–8 shows the relationship of this current to the wavelength of light incident upon the junction for the particular photodiode used in the Optec, Inc. photometer described later.

The small size of the active area on the photodiode requires that the photometer head be precisely machined and aligned on an optical bench. While quite rugged once aligned, construction of such heads is not normally attempted by amateurs.

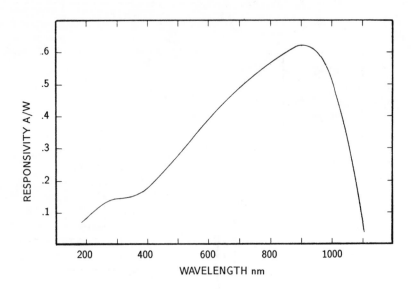

**Figure 4–8.** *Current versus wavelength for a blue-enhanced PIN photodiode. Note the extended response into the red and near infrared.*

## 4.11   Cold Boxes

Under completely dark conditions, an ideal photodetector would have no output, i.e., no current or pulses. Real photodetectors depart from this ideal and produce some output even when in complete darkness. This output is often referred to as the "dark current" or "dark count." The dark current usually can be reduced by cooling the photodetector, and cold boxes are used for this purpose. While cold boxes work well in doing their primary job of reducing dark current, they introduce other problems. The most serious one is dewing or frosting. In a well-designed cold box, dewing is overcome by making the cold box absolutely airtight, bringing the light in through an evacuated double window, and heating the outer surface of the double window with a "dew ring." Because of the difficulties encountered with cold boxes, some installations do not use them. For many small observatories a cold box may be more trouble than it is worth and there are other effective ways of reducing dark current. One way is using an end-on photomultiplier with a small effective photocathode area, although even this is not necessary unless faint stars are being observed. Small diameter photomultipliers with bialkali photocathodes can have very low dark currents if they are selected for this characteristic. On larger photomultipliers, magnetic rings can create an effectively small photocathode.

If one is trying to work to faint light levels in the near infrared, however, then cooling the photodetector becomes a necessity if it is a PMT. Cold boxes fall into two general categories, depending on the method of cooling. One type depends on the addition of the cooling medium, usually dry ice, while the other uses a built-in thermoelectric cooler. Cold boxes using dry ice are generally easier to build. Alcohol is sometimes added to the dry ice to increase thermal conductivity. When one is not trying to reduce the dark current to extremely low levels but wants merely to bring it down to a reasonable value on hot days and stabilize the temperature of the photomultiplier, then the use of $H_2O$ ice in water is cheap and convenient. It is not so likely to produce dewing, although moisture would be introduced unless the system were absolutely watertight.

Because the spectral response of some photomultipliers is a slight function of temperature, thermal stability can help minimize certain systematic errors related to color response. A stable temperature also helps to stabalize the zero point of the system. While a highly stable zero point is not required for differential photometry, it is required for all-sky photometry, which can be a very helpful system check.

Thermoelectric (TE) coolers fall into two general categories depending on how the heat is removed. One method of heat removal uses air at ambient temperature; usually air is blown over a set of fins with a fan. The other is to remove the heat with a fluid, usually water. The water can be ordinary tap water or it can be water continuously re-circulated through the thermoelectric cooler and an external heat exchanger. TE coolers not only reduce any dark current but, if left running continuously, can provide a highly stable environment for the PMT that is conducive to long life and accurate measurements.

## 4.12 Physical Considerations

In designing a photometer head, careful consideration should be given to how the photometer is attached to the telescope. For very small lightweight photometers, such as those used on many small Newtonian telescopes, the head is simply "hung" from the eyepiece holder, although it is a good idea to provide additional support, perhaps by means of a rod and sleeve with a clamp. Larger heads require a more solid mounting. A good approach to use with large Cassegrain telescopes is to have a plate on the end of the photometer and bolt that plate to the end of the telescope, i.e., to its tail stock. Whichever approach is taken, the photometer should be solid and maintain optical alignment, yet still be fairly easy to remove.

A star viewed through the microscope should not move as the photometer head is jiggled. If the secondary mirror in the main telescope (in Cassegrain systems) is not easily adjustable for focus and the photometer is a large one or uses a cold box, then considerable thought is required to make focusing convenient without compromising rigidity.

If use of a cold box is contemplated or if several different exchangeable photo-multiplier containers are to be used, then additional thought is required to decide how the photomultiplier box or cold box will be attached to the main part of the photometer head. In any case, the main head must be strong enough to support the load or else must be strengthened with external supports.

It is of utmost importance that the photometer exclude all light except that from the intended star and sky background. Some photometrists prefer to operate with a fairly high level of red work light in the observatory so that charts can be read, data logs can be filled out, etc. It is imperative that this light not make its way into the photometer. It is relatively easy to check for light tightness by shining a flashlight around all the surfaces and cracks of the photometer head. If you do this, use an old photomultiplier tube or at least turn the high voltage down and proceed carefully. While black tape is used often as the remedy for light leaks, it is preferable to avoid them in the first place. In general, simple butt joints are insufficient to keep the light out and thus must be covered. A more complex joint is better. In general, light can make it around one corner, has a tough time with two, and cannot make it around three. Note that, if R and I measures are contemplated, a red work light will not do; moreover, note that most black tape is rather transparent in the infrared.

In an effort to avoid light leaks, do not deprive yourself of easy access to the interior of the photometer head. A generously sized cover or two for easy access to the interior of the head is most appropriate, because something is bound to go wrong sometime. A well-designed seat for the covers (a recessed or U-shaped lip) will not leak light. Felt is used sometimes with a cover to help seal it.

General access is needed to the photomultiplier box also or, alternatively, one should be able to remove the entire photomultiplier and socket from the box. It is important that no light leak in here!

Thought should be given to how the various optical parts are to be aligned and adjusted in the photometer head. With the photodetector removed and the telescope pointed at a bright object such as the moon, a small piece of paper can be inserted into the head with the cover off and a check made to see if the light is passing through the center of the lenses, mirrors, diaphragms, and filters. While small, highly compact heads have their advantages, larger roomy heads are generally easier to adjust.

During head design very careful thought should be given to making the photometer easy to operate late on some very cold night with poor lighting. The eyepieces should be placed for easy viewing. Levers and knobs should be large and easily grasped while wearing gloves. Detents and springs should have positive action. Large, easily read labels should indicate the functions and positions of the controls. Keep in mind that almost all of the controls will be used many thousands of times in a single observing season. It is perhaps most important that the controls not require so much force to operate that their movement will displace the star from the diaphragm.

## 4.13   Considerations in Selecting a Photometer Head

There are three main options to consider: buying a commercial head, making your own from a proven design, and designing and building your own from scratch. Some proven commercial and homebuilt units are described later in this chapter; even if you intend to design your own, it is worthwhile to review the features and approaches taken in these designs. In making a selection, it is important to consider not only cost but also your own expertise and skill as well as that of reliable friends or co-workers.

You also must decide what type of photodetector will be used. While the venerable photomultiplier, as exemplified by the still popular 1P21 and various EMI Gencom end-on photomultipliers, is used most often, photodiodes have been used successfully also.

Consideration must be given also to the telescope upon which the head will be mounted, particularly if the telescope is already a "given." Fork mounts usually have limited clearance. Newtonian telescopes with small $f$–ratios generally will not work with heads designed for large $f$–ratio Cassegrain telescopes. The total instrument weight which a given telescope can handle must be kept in mind.

Finally, the intended type of observing must be considered. Will the telescope be used for photoelectric photometry exclusively or only occasionally? Will the observing be with only one filter or no filter, or will many different filters be required? Will a cold box be added later? It is very difficult make the many, perhaps hundreds of decisions required in putting together a complete photometric system. While this book is intended to facilitate this process somewhat, there is no substitute for some observing experience. For this reason, we suggest that you begin with a photometric system already build by someone else or, if that is not possible, listen to the advice of experienced photometrists. One of the old-timers once remarked, "By the time I had designed and built my third system and gone through several years of observing, I really began to get the hang of it."

## 4.14 Examples of Photometer Heads

Many excellent photometer heads have been designed and built, and it is unfortunate that all of them cannot be included. An attempt has been made to give a representative sample of several relatively simple photometer heads.

The head designed by Arthur J. Stokes is one of the easiest to build. In fact, it was designed specifically to be built by amateurs with ordinary hand tools. Stokes has been building heads for over 20 years, and his design has evolved in the direction of increasing simplicity and ease of construction.

The head designed by William H. Allen neatly solves the difficult problem of clearing the fork on the popular Celestron–8 telescope. As Allen points out, it would be a shame to be the southernmost amateur equipped for photoelectric photometry and be unable to view the sky around the celestial South Pole. Allen recently has moved to Blenheim, farther north on the South Island than Dunedin, and now uses a different photometer with a 12–inch telescope in a well-appointed permenent observatory. His design for the C–8, however, remains a classic well worth studying.

Another head design for the popular Celestron–8 telescope is the one by Robert Cadmus, Jr. While it does not quite clear the fork, as the Allen head does, it includes a microscope (postviewer).

Two commercially manufactured units employing photomultiplier and photodiode detectors, respectively, are described. These units are rugged, trouble-free, relatively low in cost, and thus a good choice for those who do not wish to construct their own heads.

Finally, a head designed by Jeffrey Hopkins is described. It contains some of the best features of the Stokes and Allen heads, is easily built, and nicely clears the fork on the Celestron C–8 and similar telescopes.

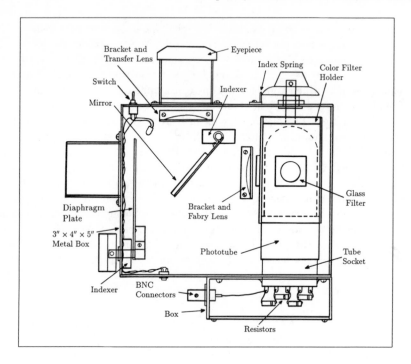

**Figure 4–9.** *Side view of the compact Stokes head.*

### 4.14.1    A Simple Photometer Head for Home Construction [5]

The lightweight compact photometer head shown in Figs. 4–9 and 4–10 can be built with hand tools and generally-available parts. The entire unit is assembled into a $3 \times 4 \times 5$–inch metal box (BUD AU–1028HG). A smaller plastic box (Radio Shack 270–230) is used to mount the photomultiplier tube, eleven-pin socket, voltage divider resistors, and BNC connectors for the high voltage and signal connections. A $1\frac{1}{4}$-inch hole cut into the bottom of the metal box allows the photomultiplier tube subassembly to be removed to change phototubes.

The box was made light-tight by putting epoxy cement on the inside of all the corner joints and then painting the inside with flat black paint. Thin cardboard gaskets were used between the two side plates and the box.

A rectangular metal turret assembly holds the three sets of glass filters for UBV photometry. The turret rotates concentrically around the phototube in a single $\frac{1}{4}$-inch bearing mounted in the top of the metal box. Old potentiometers are a good source for $\frac{1}{4}$-inch bearings. Pieces of filter glass are glued in place over the $\frac{1}{2}$-inch holes. A fourth hole can be made for unfiltered observations if desired.

The Fabry lens, eyepiece transfer lens, and movable mirror are all mounted on the right-hand side plate of the metal box. The complete plate can be removed by taking out four screws. The left-hand side plate has no parts mounted on it.

---

[5] Written by Arthur J. Stokes

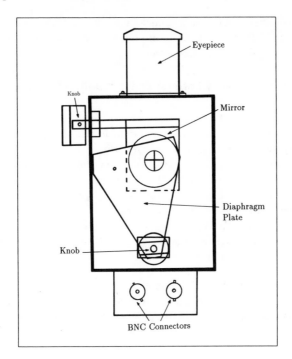

**Figure 4–10.** *End view of the Stokes head.*

The diaphragm plate is mounted at the front of the box on a rotary index assembly taken from a 12–position selector switch. This provides an easy answer to 30° indexing of the diaphragm plate. A similar assembly is used to flip the mirror up and down 30°. The diaphragm plate has a $\frac{1}{2}$–inch hole with crosshairs (0.005–inch wire) for centering on the star and a $\frac{1}{32}$–inch hole for measurement of star brightness. Crosshairs are glued in place.

The eyepiece holder was made from a piece of $1\frac{1}{4}$–inch inside diameter brass tubing soldered to a thin annular piece of sheet brass. A similar unit made from $1\frac{1}{4}$–inch outside diameter tubing was used to hold the photometer head in the telescope eyepiece holder.

The eleven-pin photomultiplier tube socket was cemented into a $1\frac{1}{8}$–inch hole cut into the top of a $3\frac{1}{4} \times 2\frac{1}{8} \times 1\frac{1}{8}$–inch plastic box. The socket must be positioned so that the photocathode faces toward front of photometer head. The two BNC connectors were mounted in $\frac{3}{8}$–inch holes in the front of the box. The 0.25–watt resistors for the voltage divider were soldered directly on the socket terminals. All electrical terminals should be cleaned of flux after soldering to prevent current leakage. Three short sheet-metal screws were used to fasten the plastic box to the bottom of the metal box. A small circular gasket cut from felt or thin foam rubber sheet placed around the socket will make a light-tight joint.

A small, single-pole, single-throw toggle switch was placed on the top of the metal box to control the light-emitting diode (LED) used to illuminate the crosshairs and diaphragm hole during centering of the star. The LED was soldered to the switch terminals with a 2200 ohm resistor in series with one lead wire. A pair of wires was run to a miniature audio jack mounted on the bottom of the metal box. A 9 or 12 volt DC wall-plug-type supply similar to Radio Shack 270-1552 can be used for the LED power.

### 4.14.2  A Photoelectric Photometer for Small Telescopes [6]

#### Introduction

For many years I have been involved with photoelectric photometry at the Beverly Begg Observatory in Dunedin, New Zealand, where we have a 30 cm Cassegrain telescope, an automated photoelectric photometer, and a high-speed data logging system. I have felt, however, that there is a real need to develop a compact portable photometer which can be fitted to a small telescope and permit useful photoelectric work to be done at small observatories or in the backyard at home.

Having just purchased a Celestron–8 telescope, I decided to use this instrument as a basis upon which to define the design parameters for the photometer. The following design parameters were adopted:

1. The photometer head must be small enough to pass between the forks, thus giving access to the polar regions. A friend of mine once remarked that it is no use having the most southerly photoelectric system in the world without being able to observe the southern polar sky.

2. The diaphragm size is to be chosen within the limits of the tracking accuracy of the C–8 sidereal drive.

3. In order to minimize disturbance of the telescope once it is set on a star, the filter wheel and data readout system should be automated.

4. All of the necessary control and data output electronics should be mounted in the photometer head. This will reduce the number of auxiliary cables between the photometer head and the output instrumentation.

#### Optical and Mechanical Design

To meet the design criteria detailed above, the photometer head must be very compact. To achieve this, the photometer has a 90° two-axis flip mirror, which eliminates the need for the two flip mirrors in a conventional photometer head, and a drum filter wheel on which the three filters are mounted.

The shell of the photometer head, shown in Figs. 4–11 and 4–12, is constructed from aluminum rectangular of hollow section. The various sections are fabricated with aluminum flanges and can be bolted together in three sections.

Light from the telescope passes into the photometer head as shown and strikes a movable aluminized glass mirror. This and the other optical components were obtained from the Edmund Scientific Company. This mirror is mounted on a machined brass holder which can move the mirror in two directions. In the first position, it directs the light past the illuminated crosshairs

---

[6] Written by William H. Allen

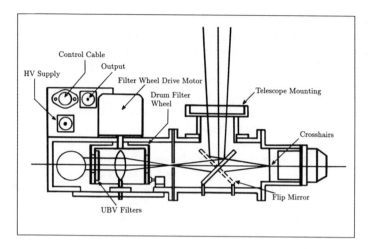

**Figure 4–11.** *Side view of the Allen photometer head. This highly compact head clears the fork on Allen's Celestron C–8 telescope, allowing him to access the South Celestial Pole.*

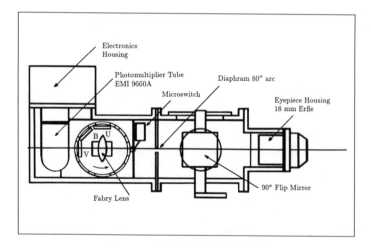

**Figure 4–12.** *Top view of the Allen photometer head. The placement of the filters on the filter wheel is easy to see.*

and into the wide angle 18 mm Erfle eyepiece; in the other position, 180° away, it directs the light into the photoelectric measuring section.

The mirror is so arranged to rotate through 90°, the plane of rotation being parallel and on the aluminum surface. This allows the optical axis to be axial in both directions, thereby simplifying alignment of the instrument. The adjusting screws shown permit precise optical alignment.

The flip mirror section of the instrument is made symmetrical, so that the telescope focal plane can be placed either at the crosshairs or on the diaphragm plate. This lets the crosshairs be set up in a position which exactly represents the position of the diaphragm and reduces the need for the second diaphragm-viewing telescope used in most conventional photometer heads.

The tracking accuracy of any small telescope with mass-produced sidereal gears is limited for photometric observations of long duration because of the difficulty of keeping the star in the diaphragm. For a Celestron–8 telescope, I have found by experiment that the diaphragm should be no smaller than about 80 arcseconds in diameter. Such a size lets a three-color observation be made with little chance of the star moving out of the diaphragm.

After the light passes through the diaphragm, it passes through a Fabry lens. The purpose of this lens is to image the spot of light on the photomultiplier photocathode in such a way that movement of the star image does not cause a variable output from the photomultiplier because of variable sensitivity across the photocathode surface. The starlight evenly illuminates the telescope objective, which is imaged by the Fabry lens as a spot of light on the photocathode. Since the telescope objective, Fabry lens, and photocathode are fixed within the photometer head, any movement of the star image in the focal plane will not cause the spot of light on the photocathode to move.

The placement of the filters on the filter wheel is easy to see. The parallel rays of light now pass through one of the three filters mounted on the inside circumference of the drum. The drum is driven by a 12V geared electric motor and is automatically stopped at each filter position by a microswitch which opens the 12V supply to the motor. A DC motor will run on after its supply is removed and thus would leave the filter placed not exactly in the center of the light beam. Because of small variations in filter transparency, it is essential that the light pass through the same part of the filter each time, to ensure consistency in measurement. To stop this overshoot, we use regenerative braking to stop the DC motor instantly.

The drum filter wheel is controlled remotely from a hand-held control unit. The crosshairs are illuminated by a red light-emitting diode, which is adjusted in brightness also from the hand-held control unit.

The drum filter wheel is moved by a geared DC motor, the gearing of which is chosen to give a two-second filter change period for each of the four positions. Due to the design of the drum filter, which requires twice as many holes as filter positions, it takes eight seconds to reset the start of the sequence (no filter, V filter, B filter, U filter). This delay is not a problem because it provides a method of indexing the drum, thereby enabling the observer to know exactly which filter is being used at the start of each cycle. It also eliminates the need for expensive readout systems to show the filter position; one need only listen for the long run of the motor which, when stopped, indicates the start of the sequence. With practice, the observer readily becomes attuned to this eight-second motor run.

### Setting Up the Instrument

Because no diaphragm inspection eyepiece is used in this instrument, it must be aligned accurately so that, when a star is positioned on the crosshairs of the field eyepiece, it will also be positioned in the center of the diaphragm when the mirror is flipped over to the other position.

This is done in two stages, with the flip mirror adjustment screws. First, the photomultiplier tube is removed and the optical axis aligned by viewing the spot of light through a hole in the end of the housing. The spot of light is moved by adjusting the flip mirror position so that the optical and mechanical axes are aligned. Second, with the optical axis thus aligned in one plane, it is aligned in the other plane by moving the photomultiplier tube in or out. Because of the rectangular photocathode in the EMI 9660A tube which is used in this instrument, the second alignment need not be so precise.

Adjustment of the optical axis in the other direction is made by setting a star in the diaphragm and adjusting the flip mirror to center the star on the crosshairs. Again the star can be positioned only in one plane by this method and movement of the crosshairs in the other plane enables the star to be centered on both crosshairs. In practice it has been found that, once the instrument is set up, it is very easy to position the star in the diaphragm without actually viewing it.

## Operation

The object to be measured is first located in the field eyepiece and moved into the center of the crosshairs. The flip mirror then is rotated through 90°, the star will now be in the center of the diaphragm, and the photoelectric observations can commence. If, however, one wishes to verify the centering, one can do this remotely by moving the telescope right ascension and declination slow motions and observing the change in output on the frequency counter.

This is best done by setting the frequency counter sampling time to a one-second interval. When the star moves out of or near to the edge of the diaphragm, there will be a sharp drop in count rate. With experience one can position the star accurately in the center of the diaphragm without viewing it. If there are verniers on the right ascension and declination circles, direct positioning can be made by reference to the vernier scale.

When the filter wheel position is changed remotely, it will be indexed in a regular way, i.e., no filter, V filter, B filter, and U filter. An observation sheet can be drawn up to record the observations conveniently with this indexing in mind.

## Uses of the Photometer

The Celestron–8 telescope with this photometer is very compact and is regularly used in my backyard to observe variable stars. Because of the photometric system's small size, it does not take long to set up and does not require a permanent observatory. It does, however, require an accurately aligned pier to permit accurate tracking. This pier and the adjustment screws are shown in Photo. 4–1, while a close-up of the various components of the entire system are shown Photo. 4–2. The electronics portion of the photometer head can be seen in Photo. 4–3.

I also use this instrument at the Mt. John University Observatory as a sky monitoring photometer when I observe flare stars with the main 24–inch instrument. I observe the flare star itself with the 24–inch telescope and its

**Photograph 4–1. (Left)** *Allen photometer and Celestron C–8 telescope shown on the permanent pier at his home in Dunedin, New Zealand.*

**Photograph 4–2. (Right)***The entire Allen photometric system. Shown are the high voltage power supply, frequency counter, hand control unit, and telescope.*

**Photograph 4–3.** *A view of the Allen photometer head detailing the current-to-frequency converter and photomultiplier dynode resistor chain.*

photometric equipment and observe a nearby comparison star with the 8-inch guide refractor on which is mounted my small photometer.

### 4.14.3 The Grinnell College Photometer[7]

In designing a photometer for a small telescope such as the Celestron 8, one faces two problems which are usually not significant when one is dealing with a larger telescope. The first problem is size and weight. Although counterweights will be required in any case, keeping the weight of the photometer low and keeping its center of gravity close to the telescope allow the instrument to be

---

[7]Written by Robert R. Cadmus, Jr.

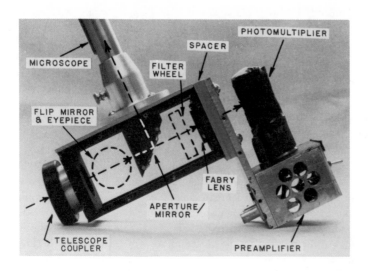

**Photograph 4–4.** *Internal structure of the Grinnell photometer head. The combination aperture (diaphragm) and microscope mirror is a unique feature of this photometer.*

**Photograph 4–5.** *External view of the Grinnell photometer. It is shown attached to Grinnell College Celestron C–8 telescope.*

supported entirely by the threads at the rear of the telescope, without additional brackets. This arrangement makes it easy to rotate the photometer about the optical axis and simplifies the installation and removal of the device.

The second problem is that of keeping the star in the diaphragm. Even when a small telescope is permanently and rigidly mounted, there can be a considerable periodic drift in right ascension. This effect, compounded by image motion due to wind or other mechanical disturbances, reduces one's confidence that the star remains centered in the diaphragm after the flip mirror or microscope has been withdrawn. In the Grinnell College photometer, which is described below, this problem has been solved by using a diaphragm which is not an actual hole but rather is an unaluminized spot on a mirror which is tilted at 45° to the optical axis. The stellar image is formed at the diaphragm so that, when one views the diaphragm from the side with a microscope, both the star and the diaphragm are in focus (See Photo. 4–4). The field seen through the microscope is just a normal high-power image of the sky with a "hole" in the center corresponding to the diaphragm. An illuminated reticle in the microscope tells you where the center of the diaphragm is when the sky is too dark for the diaphragm itself to be visible as a spot. If the star wanders out of the diaphragm, it immediately becomes visible in the field, although the measurement must be discarded at that point anyway. Because the unaluminized spot of the glass reflects about 8% of the incident light, stars brighter than about 10th magnitude are visible even when they are in the diaphragm. The observer can monitor the star's position in the diaphragm and make any necessary corrections while data are being taken with the photometer.

The Grinnell College photometer was designed following conventional practice except for the use of this beam-splitting mirror. Although weight was a concern, the instrument was made to be stiff and strong. In addition, some increase in weight was tolerated in order to make the photometer modular, in the sense that the subassemblies are attached to flat plates which can be removed individually and modified easily. The internal structure of the photometer is shown in Photo. 4–4 and the external appearance is shown in Photo. 4–5.

The photometer is attached to the telescope with a heftier version of the standard Celestron clamping-ring arrangement. The front plate of the photometer is thick because it must support everything behind it. It is attached to the rear plate by four square rods which serve primarily to tie the edges of the side plates together in order to reduce light leaks. Behind the front plate is a flip mirror feeding a regular telescope eyepiece. In the Grinnell design this is used only as a "finder" and not to center the star in the diaphragm. It also permits normal visual use of the telescope without removing the photometer.

The diaphragm is attached to the beveled end of the microscope holder assembly so that the microscope and diaphragm can be removed from the photometer as a unit for alignment. The glass diaphragms are actually mounted on accurately machined aluminum blocks which fit in a recess on the bottom surface of the microscope holder. Thus, a number of different diaphragms can be pre-aligned on separate blocks and easily interchanged. In practice the diaphragm size is not frequently changed, because the minimum usable diaphragm diameter is determined by the precision of the tracking and slow motion controls of the telescope.

Following the diaphragm is a six-position filter wheel and, following that, the Fabry lens mounted in the rear plate. The photomultiplier tube housing is a separate unit which is attached to the rear plate of the photometer by means of a hollow spacer. The spacer permits a larger housing to be installed later, without moving the photocathode back. The housing contains a preamplifier for use in making high-speed measurements with an on-line computer. Because the tube housing is the part farthest from the telescope, excess weight there puts a particularly large stress on the telescope coupler. For this reason, the housing was made as light as possible by drilling holes and milling away material in low-stress areas.

The photometer was constructed almost entirely of aluminum and weighs 1.73 kg (3 lbs. 13 oz.). Its center of gravity is 11 cm (4.75 in.) behind the end of the telescope threads and the package is handled well by either a Celestron-8 or a Celestron-14. The diaphragms are made by vacuum evaporation of aluminum onto 1 mm thick microscope slides, with small plastic beads shadowing the aperture areas. Elliptical apertures, which appear circular both to the incoming starlight and through the microscope, can be produced by tilting the slide relative to the source during evaporation. If vacuum evaporation facilities are not available, it might be possible to remove a spot of aluminum from a mirror chemically, but getting a sharp-edged diaphragm in this way may be difficult. The diaphragm commonly used at Grinnell is about 0.6 mm in diameter, corresponding to about 70 arcseconds on the sky. The 1 mm-thick glass works well because light reflected from the rear surface cannot pass back through a typical diaphragm and produce a second image. It is not thick enough, however, to produce a significant shift of the light beam.

### 4.14.4 Solid-State Stellar Photometers [8]

#### Introduction

The photometer described in this section is a state-of-the-art electronic photometer designed exclusively for stellar photometry. It is the first stellar photometer for general astronomy use that employs a blue-enhanced silicon PIN photodiode detector instead of the usual photomultiplier tube. The use of this detector permits measurement from the ultraviolet to the infrared.

The system, shown attached to a Celestron C-8 telescope in Photo. 4-6, is rugged and simple to use. Simply place the lightweight, all-metal telescope coupler into the telescope's focusing tube, center the celestial object within the illuminated ring of the focusing eyepiece, flip the viewing mirror out of the optical path, and observe the output on the large illuminated meter. A 100,000/1 light range is observable with the gain selector. Almost indestructible, the silicon detector cannot be damaged by accidental exposure to bright lights or rough handling.

The telescope coupler houses the silicon detector, electrometer, and focusing eyepiece and fits standard $1\frac{1}{4}$-inch focusing mounts. Precision-machined from 6061-T6 aluminum alloy, the unit is anodized black with a special brightening

---

[8]Written by Gerald Persha

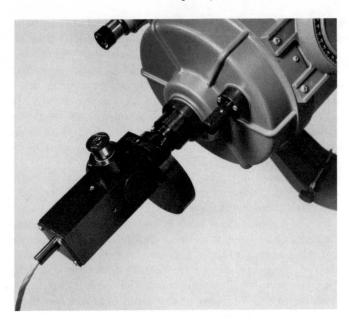

**Photograph 4–6.** *Optec solid-state photometer attached to a Celestron C–8 telescope. The turret can hold six filters.*

process for added corrosion resistance.

## Detector

The state-of-the-art, blue-enhanced silicon PIN photodiode is specifically manufactured for this photometer. The photodiode functions in the photovoltaic mode and features a noise equivalent power (NEP) of less than $7 \times 10^{-15} W/Hz^{1/2}$ and a circular sensitive area 0.5 mm in diameter. In the photovoltaic mode, this detector has no dark current and responds linearly (less than 1% error) with light to the saturation point (10V output) of the electrometer amplifier. Low light detection is limited by thermal noise generated in the P-N junction of the photodiode. The noise current $I_n$ is given by

$$I_n^2 = \frac{4kTB}{R},$$
(4.14.1)

where $k$ is the Boltzmann constant and $T$ is the temperature in degrees Kelvin.

The detector used in this photometer has a shunt resistance (R) greater than 1000 M ohms and is the highest now available. Noise current is further reduced by limiting the bandwidth (B) to 0.1 Hz when operating the photometer in the slow response mode. The response of the detector is 0.47 A/W at 850 nm and, through blue-enhancement during the manufacturing process, is nearly 0.10 A/W at 360 nm. The detector's spectral response is shown in Fig. 4–13.

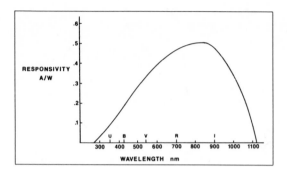

**Figure 4–13.** *Spectral response of the solid-state detector. The response extends into the far red and near infrared.*

## Focusing eyepiece

The 1–inch F. L. Ramsden eyepiece has an apparent field of view of 35°. A reticle, with an inscribed ring in the center, coincides exactly with the position and size of the detector's sensitive area. With illumination from the side by a long-lasting red LED, the inscribed ring stands out brightly against a dark background.

## Operation

Professional and amateur astronomers can make accurate photometric observations with this photometer. The procedure used, for example in the case of variable stars, is to select the object to be measured and a nearby comparison star of similar color and known magnitude. The coupler is attached to the telescope and the observer then selects a region of the sky with no stars, flips up the mirror, and zeros the meter. The observer then lines up the program object and makes a reading. After recording the first reading, he makes a comparison reading of the comparison star.

## Filters

The optional filters and filter turret allow photometric observations to be made at one or more specific wavelengths. This lets the observer do work in such fields as infrared astronomy and colorimetry (the measurement of stellar temperatures). Filters are available to match the standard Johnson UBVRI system and are factory-installed in the filter wheel. The customer has the choice of any combination of filters. The filter wheels for the turret are interchangeable, with 4–, 5–, or 6–position models available. A clear window position is standard with all filter wheels. The spectral response of the UBVRI filters is shown in Fig. 4–14.

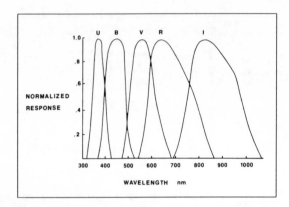

**Figure 4–14.** *Spectral response of the UBVRI filters and photodetector.*

## Assembled head

A cross-sectional drawing of the entire head is shown in Fig. 4–15. Unlike a conventional head, the filter precedes the previewer, there is no postviewer, and the diaphragm, which can be ordered in two fixed sizes, is a mask on the photodiode.

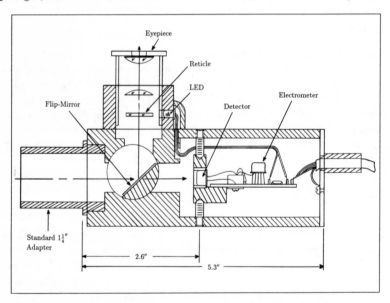

**Figure 4–15.** *Cross-sectional view of the Optec solid-state photometer. Use of a photodiode detector allows a highly compact design.*

### 4.14.5 A Photon-Counting Stellar Photometer [9]

Thorn EMI Gencom, Inc. has been a major manufacturer of photomultiplier tubes (PMT's) and PMT accessories for many years. By incorporating our considerable PMT experience and electronics expertise with the invaluable recommendations of Russ Genet and other IAPPP members, we have designed a research-grade pulse-counting stellar photometer which can be fully utilized by student, amateur, and professional astronomers.

The STARLIGHT–1 stellar photometer consists of the lightweight photometer head described in this chapter and an electronics unit described in the following chapter. The entire system is shown in Photo. 4–7.

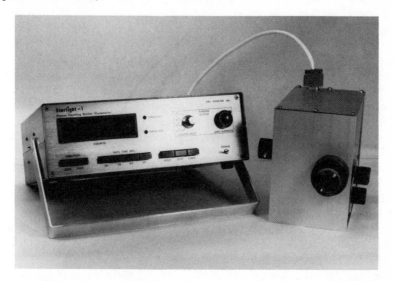

**Photograph 4–7.** *Thorn EMI Gencom, Inc. STARLIGHT–1 photon-counting stellar photometer. Shown on the photometer head from left to right are the telescope coupling adapter, viewing eyepiece, filter wheel dial, diaphragm dial, and prism position switch.*

We wanted to design the STARLIGHT–1 so that the detector would not be the limiting factor in measurement accuracy. This required a photomultiplier tube which had the following characteristics:

1. Greater than 30% quantum efficiency at peak wavelength

2. Extremely low background noise

3. A gain sufficient for single photoelectron detection

4. Small dimensions

5. UBV capability without requiring cooling

---

[9]Written by Robert C. Wolpert

**6.** Interchangeability with other PMT's mechanically identical but of different spectral sensitivities

**7.** Operate at less than 1000V DC

**8.** Low in cost.

The obvious choice was the end-window EMI 9924A. It has a RbCs photo-cathode with superb UBV sensitivity, has a dark current of less than 50 pico amps, has an excellent plateau curve at less than 200 c.p.s., is only $1\frac{1}{8}$ inches in diameter, has many spectral response variants which are mechanically iden-tical, and (last, but not least) is low in cost. This photomultiplier tube, when coupled to a fast high-gain differential amplifier/discriminator, will provide sin-gle photoelectron detection capability. In addition to the EMI9924A PMT, the detector assembly consists of a 500–1600 volt DC power supply and a $2 \times 4$–inch circuit board which contains the PMT voltage divider network, a LeCroy MVL 100 amplifier/discriminator, and a gain control pot. The power supply provides protection for the photomultiplier tube by means of current limiting with voltage foldback. Maximum output current is approximately 350 $\mu$A.

The photomultiplier tube is housed in a foamed chamber with RFI and mag-netic shielding and is designed to allow easy replacement of the photomultiplier tube.

All of the above items comprise about 40 percent of the available volume in the $6.5 \times 4.5 \times 7.2$–inch photometer head. The remainder of the available space is completely filled by a complex optical assembly.

The optical assembly was the most difficult part of the STARLIGHT–1 to design. Twice the design had to be scrapped as it neared completion in order to incorporate some newly suggested feature. The final design incorporates every desirable feature that was suggested by experienced photometrists.

Incoming light will first pass through one of the diaphragms in a 6–position aluminum diaphragm turret. Each position is detented for positive and accurate diaphragm location. Each diaphragm position also has an offset detent which will allow sky background measurements to be taken without requiring the entire telescope to be repositioned. There is a full open $\frac{3}{4}$–inch diaphragm which contains an L.E.D.-illuminated reticle. The L.E.D. is located in the flip prism mechanism and will light only when the prism is in the "viewing" position. The intensity of the L.E.D. is controlled by an external on/off brightness control dial. The same L.E.D. also backlights each diaphragm plate, allowing the observer to distinguish easily between the sky field and the edges of the diaphragm.

After passing through the diaphragm, the signal can be either reflected into the viewing microscope by a flip prism or allowed to pass through one of the filters in a 6-position aluminum filter turret. Each position on the filter turret is also accurately detented for positive location. Ultraviolet, blue, and visual filters are provided to match the UBV system. These are specially corrected for the EMI 9924A. An opaque filter position serves as a shutter which protects the PMT from ambient light while the photometer is not in use. The two unused filter positions can be utilized for red, infrared, or other user-selected filters. After light passes through the desired filter, it will pass through a Fabry lens and onto the PMT window. The entire optical system is designed so that there is no light loss in telescopes of $f/5.0$ or longer focal length.

All metal surfaces in the photometer head have been either anodized or painted black, and great care has been taken to prevent light leaks. The four control dials are conveniently located on a single outside panel. The diaphragm and filter turrets have illuminated position indicators on their dials which allow positive position identification without the need to turn on observatory lights. The third dial controls the L.E.D. illumination and the fourth will flip the prism into the desired position. A front view of the photometer is shown in Photograph 4–8. The telescope adapter tube has been designed to support the entire weight of the photometer head. A detented ring has been machined into the adapter, so that secure positioning can be provided with a set-screw. The photometer head is thus able to be mounted within seconds.

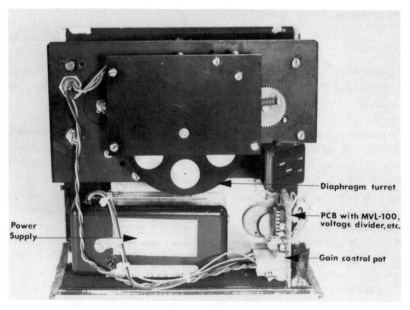

**Photograph 4–8.** *Front view of the STARLIGHT–1 with the cover removed.*

### 4.14.6   A Highly Compact Photometer Head [10]

### Introduction

The Celestron C–8 is undoubtedly one of the world's most popular telescopes. It is also an excellent choice for a small observatory PEP system. It is compact, easy to set up and use, and reasonably priced. A problem arises when buying or making a PEP head for the C–8. Because of the C–8's compact size, it requires a fairly small and lightweight PEP head.

---

[10]Written by Jeffrey L. Hopkins

I decided that my design should meet the following criteria:

1. The PEP head should pass through the Celestron C-8 fork mount and allow the system to be aimed at any point in the sky available to the telescope without the PEP head.

2. The PEP head should be completely light tight.

3. The weight should be kept at a minimum.

4. The PEP head should be easy to operate.

5. The PEP head should be easy to build.

6. The head should be inexpensive.

7. The design should have a built-in sky offset.

## Construction

The design is centered around the PEP head housing which uses a Hammond Manufacturing #1590R die-cast aluminum box $7\frac{1}{2} \times 4\frac{1}{3} \times 2\frac{3}{8}$–inch. The inside of the box is painted flat black and the outside is painted orange to match the Celestron.

To provide a compact design, two mirrors are used. The first mirror is a flip mirror used to divert the light from the diaphragm to the eyepiece so the star can be centered in the diaphragm. The second mirror is fixed and, when the first mirror is flipped out of the light path, the second mirror diverts the starlight through the Fabry lens and the filter, and on to the photocathode of the photomultiplier. A standard $1\frac{1}{4}$–inch chrome-plated brass tube (available at any hardware store's plumbing department) is used to couple the photometer head housing to the telescope. A nylon or aluminum plug is used to secure the tubing to the housing. A filter cage was machined out of aluminum to fit around the PMT and magnetic shield. A less expensive cage could be made using a piece of PVC tubing or square/round aluminum tubing. The filter cage has four main sides, three of which hold the UBV filters. The fourth side is left blank to act as a "dark slide" to protect the PMT from excess light. If desired, this fourth side could have a hole cut in it and allow unfiltered light to be measured, e.g., for occultation measurements. Two detents are provided by small indentations on the top of the filter cage. Two 6–32 spring-loaded detents (VLIER, S49, available at a machinist supply house) pass through the top of the housing and provide the proper detent location for each filter.

The diaphragm plate is made of 0.042–inch aluminum and has holes of three different sizes in it. The first is 0.5–inch to allow easy finding of the star and centering it. The next is approximately 120 arcseconds and the last is about 60 arcseconds. I have found it easiest to make the small holes in a thin sheet of brass and glue the brass to the diaphragm plate. That way you can experiment with different holes and not need to make a complete new diaphragm plate each time. If this method is used, then three 0.5-inch holes should be made in the plate. The best way to determine the actual angular size of a hole is to point the telescope toward a star on the equator and near the meridian. Time the star as it drifts from one edge through the center to the other edge (with the clock drive off). Multiply the number of seconds by 15 and the resulting number is the angular size of the hole in arcseconds. The diaphragm is positioned by a lever on the front of the housing near the eyepiece. Three main indentations provide centering of the three holes. Two additional indentations, on each side of the center indentation of the 60– and 120–arcsecond hole, provide an easy means of obtaining a sky offset without moving the telescope. A 6–32 spring loaded detent is threaded into the lever. A Mead #427 0.965–inch Illuminated Reticle Eyepiece (12 mm) is used to observe the star in the diaphragm. The relay lens assembly consists of a machined piece of aluminum to hold the eyepiece and relay lenses and to fasten the assembly to the right side of the housing. The relay lenses are fixed in a piece of PCV tubing which is machined to fit inside a piece of 1–inch copper tubing. No provisions have been made for cooling the PMT. For most applications cooling is not necessary and to add that feature would complicate the design considerably. There is still room left inside the housing to mount an MVL 100 for photon counting and/or a high voltage power supply for the PMT.

Shown in Figures 4–16 and 4–17 are scale drawings of the photometer head.

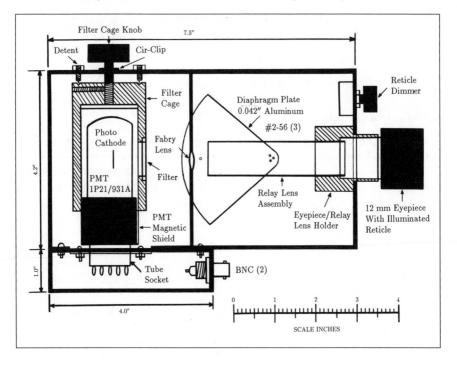

**Figure 4–16.** *Drawing of the Hopkins' photometer head as seen from the bottom.*

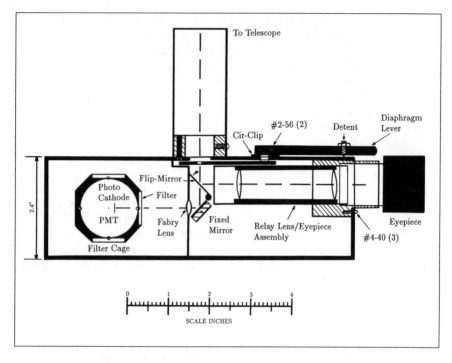

**Figure 4–17.** *Side view of the Hopkins' photometer head.*

### 4.14.7    Conclusion

The design presented here does not have to be followed exactly. Variations in materials and methods can still produce a very good photometer head. Don't depend on published sizes for lens, tubing, etc. Measure them yourself and proceed from there. It is very important to take the time to align all light paths accurately. When completed, this photometer head should provide many years of successful photoelectric photometry.

# Chapter 5

# SUPPORTING ELECTRONICS

## 5.1 Introduction

The supporting electronics supply the photodetector with the necessary voltages for proper operation (the power supply) and take the output of the photodetector and amplify it (the DC amplifier) or convert it (pulse amplifier/discriminator) to a usable form. Simple electronics used to log the data are covered in this chapter, while the use of microcomputers to log the data or control the photometric system will be covered in the following chapters.

Photomultiplier tubes require a rather high voltage (about 1000 volts DC) for proper operation. As their gain is a sensitive function of the voltage supplied, it is necessary to regulate this high voltage so that it will remain constant in spite of changes in line voltage, temperature, etc.

The starlight focused on the photocathode of a photomultiplier consists, of course, of discrete quanta of light, i.e., photons. Only a fraction of the photons result in the release of an electron from the photocathode. Each electron released from the photocathode results in approximately one million electrons arriving at the anode in a short pulse, the actual number depending on the voltage applied to the photomultiplier. When his 10–inch telescope is looking at a 5th magnitude star, Howard Louth finds that 200,000 electrons per second are released from his photocathode, resulting in 200,000 pulses per second (ignoring overlaps) of about one million electrons each. Since there are $6.25 \times 10^{18}$ electrons per coulomb and one coulomb per second is defined as an ampere, the current output is about $32 \times 10^{-9}$ amps or 32 nanoamps. Note that this is the average current output. The current from a single short pulse is much higher, as explained in the previous chapter.

Thus, the output of a photomultiplier is a series of very weak current pulses. There are two main approaches to bringing this weak signal up to a useful level. One is to treat it as a small current flow of low frequency, essentially ignoring its pulse nature, and then amplify this weak current with an operational amplifier of very high input impedance. The other approach is to amplify the weak pulses with a wideband amplifier, detect the crossing of the pulses across a fixed threshold with a discriminator, and count them. Both approaches have produced excellent results and each has its own advantages and disadvantages. These will be discussed later in this chapter.

## 5.2   High Voltage Power Supplies

### 5.2.1   Introduction

For proper operation photomultiplier tubes need to be supplied with a high DC voltage. A single voltage is usually supplied and a string of series divider resistors is used to provide the individual voltages required by each of the dynodes and the anode. The anode is usually placed at ground potential while the photocathode is the most negative, with the first dynode next to most negative, etc. The voltage used on 1P21 photomultipliers is typically in the range −700 to −1000 volts. The current drawn by the tube itself is negligible (less than one microamp), with the main load to the power supply being the divider chain (less than one milliamp) and any meter used to monitor the voltage (one milliamp typically). Thus, what is needed is a rather high voltage at very low current to a load that, for all intents and purposes, is of constant impedance.

The gain of photomultiplier tubes is a very sensitive function of the voltage applied. As a rule of thumb gain is proportional to about the 7th power of the voltage. If the applied voltage were to change during the course of an observation, an appreciable error could be introduced. While many photometric systems operate quite well on a single, fixed high voltage, it is convenient to be able to select a number of different voltages. This is because different types of photomultipliers and even different tubes of the same type require different voltages. Care should be taken in using different voltages, however, because, changes in voltage can change the color response of the tube somewhat.

Early photomultipliers typically used a number of batteries in series to provide the necessary high voltage. While this procedure worked reasonably well, it was sensitive to changes in temperature, expensive, bulky, and somewhat hazardous and therefore is not used much anymore. A simple power supply operating off the AC main line can be made by using a constant voltage transformer to feed a second step-up transformer, the output of which is then rectified and filtered. This approach is not used too often due to the cost and bulk of constant voltage transformers and the lack of precision voltage regulation. Solid-state high voltage power supplies have become increasingly popular in recent years. They are usually designed as DC inverters. A low-voltage direct current, typically 12 VDC, is used to operate an oscillator at a frequency of several kilohertz. A small transformer then steps this up to the desired high voltage, which is rectified and filtered. A feedback path is usually provided to sense changes in the output and use this to correct the output voltage of the oscillator so that a constant output voltage will be maintained. A much simpler (although less well-regulated) solid-state power supply can be made using Zener diodes. These diodes keep a nearly constant voltage across themselves in spite of varying load or overall input voltage.

The high-voltage power supply has traditionally been a source of equipment trouble in photometry and is one of the areas where a back-up unit might be given serious consideration. A number of different units are described below, each of which is quite good insofar as high-voltage power supplies go. In making a selection one has to consider whether one needs a single fixed voltage or an adjustable voltage, what degree of regulation will be required, and whether one

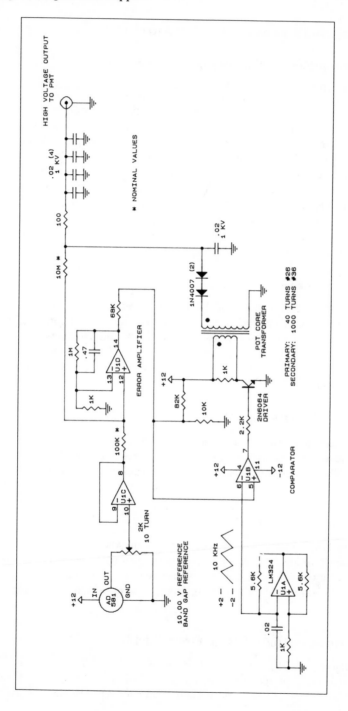

**Figure 5–1.** Louis Boyd's regulated PMT high voltage power supply.

wishes to build the power supply or buy it ready-made.

## 5.2.2   Examples of High Voltage Power Supplies

### RCA PF 1042

RCA has designed a small low-cost solid-state power supply for use with 11–pin side-on photomultipliers such as the 1P21 or EMI 9781. These devices have an integral voltage divider network and tube socket. They provide a regulated output in the range of –500 to –1200 VDC and that output is regulated to a precision of ±0.1%. They operate with an input of 12 VDC at a current of less than 100 mA. The output voltage can be set using either resistance or voltage programming (such as a simple potentiometer). They are 1.25 inches in diameter and less than 2 inches long. No high voltages are directly exposed, although great care still must be taken in the use of any photomultiplier because 1000 VDC can be lethal. At only $100 each (as of 1985), these amazing little power supplies are a good buy.

### Pacific Precision Instruments

Another commercial unit is available at a very reasonable price. While this unit does not contain an integral socket and voltage divider but rather is on printed circuit cards, it is regulated to a precision of 0.001% and has a very low drift, 0.04% in 8 hours after a 30–minute warm-up. The model 7101 operates off the AC line current and costs about $400 (as of 1985). These solid-state power supplies are of high precision.

### EMI Gencom MPS 1600 N

This commercially available precision high voltage power supply is very compact, measuring only 4.2 by 2.0 by 1.25 inches. It operates from a 12–volt input (unregulated). The voltage can be adjusted with an external potentiometer. The cost is $240 (1985).

### Boyd Regulated Supply

Louis Boyd of Phoenix, Arizona has designed the very stable high voltage power supply shown in Fig. 5–1. A voltage feedback is compared with a reference voltage generated by the AD581 bandgap reference device. Variations in the output voltage are then corrected by the error amplifier. The power supply provides excellent regulation.

### An Inexpensive High Voltage Power Supply [1]

For those people who like to build their own electronics, the inexpensive and easily constructed high voltage power supply shown in Fig. 5–2 might be considered. If the unit is battery-powered and without meter or case, the cost for parts can be under $20.

---

[1] Written by Jeffrey L. Hopkins

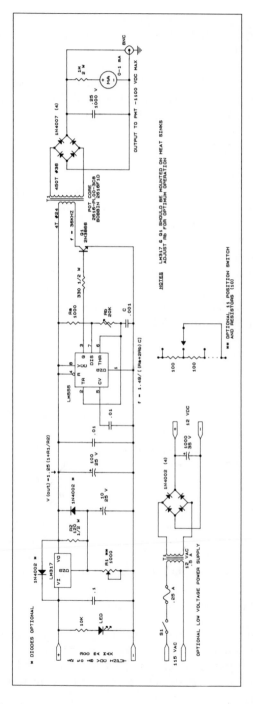

**Figure 5–2.** *Low cost pot core HVPS for PEP PMT.*

The objective of this design is to provide an adjustable, regulated power supply capable of battery operation (12 to 18 VDC) and producing 0 to –1000 VDC at 1 mA with an input current of less than 300 mA.

**The regulator.** An LM317MP three-terminal adjustable voltage regulator ($1.95) is used to provide regulated and adjustable D.C. voltage from +1.2 to +16 VDC at up to 500 mA. A 12–position wafer switch ($1.20) provides a means of selecting up to 12 different voltages. A 1K ohm 10–turn potentiometer could be used but is more expensive.

**The oscillator.** A CA555 timer ($0.59) is used to provide an approximate square wave whose amplitude is proportional to the output of the regulator. The frequency of the oscillator is determined by two resistors and a capacitor. For this design the frequency is approximately 30 kHz.

**The driver.** Almost any transistor capable of 500 mA collector current can be used. In this design an NPN type 2N3866 ($2.00) is used.

**The transformer.** The use of a pot core transformer ($6.00) greatly facilitates easy construction. Four turns of #26 enamel-coated wire are wound on the spindle first. A layer of mylar tape is then applied. After each layer a layer of mylar tape should be used. Next, 400 to 450 turns of #36 enamel-coated wire is wound with a layer of mylar tape separating each of the (approximately 5) layers. The core used in this design is a 2616+PL00–3C8.

**The output section.** Four 1N4007 diode rectifiers ($0.50/4) rated at 1000 V and 1 Amp are used as a bridge rectifier. A 0.25 $\mu$F capacitor is used to filter the output. It is made by connecting four 1.0 $\mu$F capacitors in series. The 1.0 $\mu$F capacitors are rated at 250V so, by connecting them in series, it gives a 0.25 $\mu$F 1000V capacitor ($3.20/4). A 1–megohm 2–watt resistor provides a 1 mA load and, if desired, provides scaling for 0–1000 VDC by inserting a 0 to 1 mA meter.

Although the power supply can be operated off of two 9 VDC batteries, the 300 mA load will drain them quickly. A 12 VDC car or motorcycle battery would be much better. A transformer and rectifier can provide AC operation if battery power is not used. It is recommended that the power supply be enclosed and that about 15 minutes be allowed for the voltage to stabilize.

## 5.3   DC Amplifiers vs. Pulse Amplification

Much has been written about the advantages and disadvantages of pulse counting as opposed to DC amplification in astronomical photometry. Advances in solid-state electronic integrated circuit chips have had a profound impact on both of these methods so, to a considerable extent, earlier comparisons are no longer relevant.

Traditionally, the primary advantage of DC amplifiers over pulse counters was their relative simplicity. The DC amplifier could be used directly with a relatively cheap meter or strip chart recorder and thus avoid using a high speed counter which, in the days of vacuum tubes or even discrete transistors, was complex and expensive. As another advantage, the DC amplifier also was less prone to various forms of electrical interference; its frequency response was typically below 10 Hz, while pulse counters operated with frequency responses in the megaHertz or even hundreds of megaHertz. One disadvantage until recently was that DC amplifiers were prone to drift. Another disadvantage was that their use with meters or strip chart recorders limited their overall accuracy to at best 0.1% of full scale. This limitation in accuracy in recorder resolution usually was circumvented to some extent by providing a number of ranges (different gains or amplifications) from which to select, although accurate range changing requires very careful calibration of the gain steps or amplification factors.

Theoretically, if everything is perfect, pulse counting has a $\sqrt{2}$ advantage in signal-to-noise ratio over the DC amplification approach. Of some practical significance, pulse counting has some advantages which stem from its pulse nature. Variations in the light spot on the photocathode, changes in tube voltages, and changes in tube temperature all tend to have far less effect on the number of pulses than they do on the integrated current. This benefit is offset somewhat by the temperature sensitivity of some pulse amplifier/discriminators themselves. Another advantage is that, because pulse counters do not have different ranges, calibration of ranges is not a problem. This advantage is reduced by the fact that a significant fraction of pulses are missed due to pulse coincidence when the count rate becomes high; the coincident pulse or dead-time correction needed to deal with this effect is a complication of which DC amplifiers are free. The resolution of a pulse counter is very high, being one part in a million or more, although most of this resolution is beyond any meaningful accuracy under even the best atmospheric observing conditions. Attempts to realize such high resolution require very wide bandwidths and the use of and vulnerability to problems related to UHF techniques.

The advent of integrated circuits has done much to change the traditional comparison. Operational amplifiers are available today which, for all practical purposes, have no detectable drift and near infinite input impedance. The current drift in these laser-trimmed integrated circuits is so minute that provisions to correct for or to adjust drift are not required. The linearity of these new integrated circuit op-amps also is outstanding, with 1 part in 10,000 not being unusual. Rather than losing the accuracy inherent in these new amplifiers to a much less accurate meter or strip chart recorder, they are sometimes used in conjunction with a voltage-to-frequency (V/F) converter which produces a train of pulses, the frequency of which is proportional to the input voltage with an accuracy better than 1 part in 1000. Thus, because of modern integrated circuits, the DC amplifier approach has lost some of its more serious drawbacks.

Pulse counting also has benefited from integrated circuits in the area of counting circuitry itself. The entire counting circuitry now can be contained in one inexpensive integrated circuit and counters are now generally cheaper than strip chart recorders. Furthermore, a single integrated circuit amplifier/discriminator is now available for less than $20.00.

The upshot of all this is that both approaches have gained immensely from the advantages of integrated circuits, and that very accurate results now can be obtained with either approach at a much lower cost and with greater reliability than was formerly possible. If the temperature of the PMT is nearly constant and if the high voltage does not change significantly, then the accuracy of a DC system is similar to that of a well-shielded pulse counter. Moreover, the uncertainties inherent in both approaches will be much less than the uncertainties introduced by short-term changes in atmospheric transparency in even the most outstanding observing conditions.

## 5.4  DC Amplifiers [2]

### 5.4.1  Introduction

Except for a sidereal clock, a DC amplifier is probably the most attractive component of a photometric system for in-house construction. Very few commercially built DC amplifiers offer the appropriate current input ranges, stability, digital output, and appropriate price. Furthermore, a DC amplifier generally does not require very special precautions in building technique. However, the environment for a DC amplifier for use in photoelectric photometry is relatively hostile, particularly with respect to humidity and temperature changes, and DC amplifier design must address these problems.

The requirements for conversion of photomultiplier current to final readout for astronomical photometry generally involve input currents in the $10^{-6}$ to $10^{-10}$ ampere range. We will assume throughout that the readout must be in digital form, so that a simple voltmeter is not considered adequate. Hence, an output device must have at least four digits in linear units for $0\overset{m}{.}001$ resolution. Amplifier stability should certainly aim for the $0\overset{m}{.}003$ region if possible, although differential photometry relaxes the length of time for this stability requirement from many hours to several minutes. This section will examine the various components of DC amplifier systems to meet the above requirements, and recommend specific circuits and construction hints.

### 5.4.2  DC Amplifier Principles

Before examining specific DC amplifier circuits, it is very useful to consider the basic op-amp configurations. It is taken for granted that virtually all PMT DC amplifiers will use integrated circuit operational amplifiers (op-amps), instead of discrete transistors. True, the average everyday op-amps are not good enough for this application, but in the past few years special purpose op-amps have become available at reasonable prices which are almost ideal for our application. Hence, transistor DC circuits will not be considered. There may be arguments for discrete FET transistor inputs in some cases, but even FET input op-amps are easily obtained and, in general, it is much easier to work with op-amps.

An op-amp is simply an integrated circuit made up of many transistors. It is basically a DC amplifier with almost ideal characteristics: high impedance inputs

---

[2]Written by David L. DuPuy

in the voltage amplifier mode, low-impedance input in the current-to-voltage converter mode, low output impedance, and very high gain (amplification) which is easily controllable. The gain is controlled by a feedback signal, as follows. There are two inputs, the inverting input and the non-inverting input, and the output reflects the difference in voltage on these two inputs. Another way to visualize this is that the output tries to change in such a way as to make the voltage difference on the two inputs (including the feedback voltage) zero. The op-amp gain is so large (100,000 or more) that only a tiny voltage signal is needed on the inputs in order to produce a full signal output. An op-amp also has the property that the inputs draw virtually no current (high input impedance), and this is important because it means the op-amp has little if any significant effect on the signal it is trying to measure.

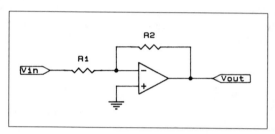

**Figure 5–3.** *Basic inverting op-amp configuration.*

We now consider a simplified analysis of op-amp operation. In Fig. 5–3 is shown the basic inverting configuration for op-amps. "Inverting" means that a positive input signal will show up on the output as a negative signal. If the non-inverting input (marked +) is at ground potential as shown, then the inverting input (marked −) will be virtually at ground potential also, because such a small signal is required to effect a change in the output. Then we see (from Ohm's law) that the voltage across $R_2$ is $V_{out}$, and the voltage across $R_1$ is $V_{in}$. Hence, $V_{in}/R_1 = -V_{out}/R_2$ since the input draws negligible current. The gain is then $V_{out}/V_{in} = -R_2/R_1$. Since the inputs are essentially at ground, the input impedance of this circuit is $R_1$, and a low input impedance may "load" the signal, changing it slightly. A "non-inverting" circuit has a gain of $V_{out}/V_{in} = 1 + R_2/R_1$, as shown in Fig. 5–4. This circuit offers the advantage of a very high input impedance: infinite impedance for the ideal op-amp, and up to $10^{12}$ ohms for an FET op-amp. These two basic configurations above can be used anywhere that a signal *voltage* is to be processed.

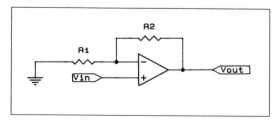

**Figure 5–4.** *Basic non-inverting op-amp configuration.*

The descriptions above assume *ideal* op-amps. In the real world, deviations from ideal devices produce slight errors. The two op-amp error specs that photometrists should be most concerned with are (a) input bias current and (b) offset voltage drift with temperature changes. The input bias current is an error current that cannot be subtracted from the signal. Hence, an op-amp must have an input bias current that is *much* smaller (say 100 times smaller) than the smallest current we wish to measure. For the ubiquitous and trusty 741 op-amp, the input bias current is 20 to 200 nanoamps (i.e. $200 \times 10^{-9}$ amperes), and that is far too large for our application. A specialized op-amp will be required.

The input offset voltage is the error voltage indicated to be on the inputs when both inputs are really grounded. This error voltage can almost always be adjusted to zero, but the problem is that this offset voltage is temperature-dependent. Further-more, even though you tweak it to zero, on a cold night it will no longer be zero and it will change during the night. Solutions to these op-amp limitations will be discussed below.

### 5.4.3   Input Stage

The signal from the PMT is a current generally in the range of $10^{-6}$ to $10^{-10}$ amperes (or 1000 to 0.1 nanoamperes). The basic op-amp configurations given above will not directly handle an input current. Hence, some device must be used to convert the signal current into a signal voltage.

The original current-to-voltage converter is the everyday resistor. Ohm's law gives the voltage across the resistor as $V = iR$. This is known as a passive current-to-voltage converter, and it is useful in many situations, but it should be avoided for use with PMT applications. The basic principle can be seen in Fig. 5–5, when the PMT current passes through R and produces a voltage across the resistor. The high input impedance of the op-amp means that essentially no PMT current passes through the op-amp. This passive current-to-voltage converter will give slight errors (Young, 1974): since the resistor is not an ideal current-to-voltage converter, it offers a resistance in the current path, which could change the current. Furthermore, it raises the PMT anode above ground. This leads to slight, but definite, errors. Finally, switching this input resistor for different gain steps offers variable loading on the PMT signal.

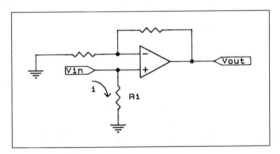

**Figure 5–5.** *Incorrect configuration for using an op-amp as a current-to-voltage converter.*

There is a simple approach that yields more accurate results. The op-amp can be wired as an active current-to-voltage converter ("active" here means that integrated circuit transistors are employed). The proper technique for measuring small currents is to use the configuration shown in Fig. 5–6. In this circuit, the op-amp offers virtually zero input resistance, so the signal current is not affected. The output voltage is $V_{out} = -i_{in}R_f$, where $R_f$ is the feedback resistor. The resistor $R_1$ is optional, but it is highly desirable to help prevent offset voltage drift with temperature changes. Ideally, the value of $R_1$ should equal that of $R_f$, but a 1-to-5 megohm resistor may be the largest practical value.

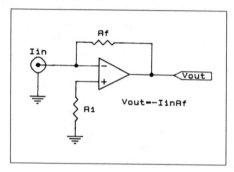

**Figure 5–6.** *Basic configuration for using an op-amp as a current-to-voltage converter.*

For the current range of interest to us, a very large value of $R_f$ is implied: if $i = 10^{-10}$A, then $R_f = 1000$ meg ohms gives only 0.1 volt output. Very large value resistors are expensive and perhaps difficult to find, and a way around this problem is to use a "gain multiplication" circuit, shown in Fig. 5–7. Only a fraction of the feedback signal is made available (via the voltage divider $R_1$ and $R_2$), and the effective gain is correspondingly higher. The multiplication is roughly $1 + R_1/R_2$, and a good approach is simply to switch $R_2$ only. Resistors for this circuit should be low-noise, thin film types. Although the Johnson noise in the resistors will have less effect for larger value resistors, this effect is completely negligible in this type of circuit for virtually any value of feedback or input resistor.

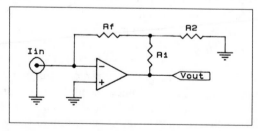

**Figure 5–7.** *Op-amp current-to-voltage converter with gain multiplication circuit to avoid using extremely large value resistors.*

Considerable care should be given to the choice of op-amp in order to meet the desired goals. Table 5–1 lists several choices of op-amps, along with specifications on input bias current and offset voltage drift. Note that the op-amps with a small voltage drift generally do not have a small input bias current; here, a compromise is required. The LF 355 is an excellent choice for experimenting, because of the cost. The AD 515J offers an astonishingly small input bias current (even smaller with the K version), but with a relatively large offset voltage drift. The 310J is a hybrid (large, encapsulated package), and is an older design. For differential photometry of faint stars, the AD 515 is a good choice, with a front panel control to re-zero the op-amp. Note that the input bias current (an error current) for an FET op-amp like those in the AD 515 or LF 355 family decreases as temperature decreases. Unfortunately, the chip temperature may be higher than ambient temperature, because of the power dissipated by the chip in normal operation, and you may not therefore realize the low input bias current shown on the spec sheet for a chip temperature of 25°C. In any case, an FET op-amp should show a smaller input bias current on very cold, clear nights. The best solution is to pick an op-amp with an input bias current in the 1 pA range or less, and a check on the linearity can be made with a constant current source if you will be trying to observe stars producing photocurrents in the 10 pA range or less (.e.g., faint red stars with the U filter). The ICL 7650 op-amp will be discussed later. For the usual astronomical photometry (i.e., telescope of average size and stars of average brightness) I do not consider an input bias current of 100 pA to be adequately small; one should use an op-amp with an input bias current of 1 to 5 pA or smaller, except perhaps for preliminary experiment with the inexpensive LF 355A or similar op-amp.

| TABLE 5–1 | | | |
|-----------|------|-------|----------|
| Device    | Bias | Drift | Cost ($) |
| CA 3140A  | 10   | 6     | 1        |
| LF355A    | 30   | 3     | 1.50     |
| BB 3528A  | 0.3  | 15    | 20       |
| AD 515JH  | 0.15 | 15    | 20       |
| OP–07A    | 700  | 0.2   | 45       |
| AD 310J   | 0.01 | 30    | 78       |
| ICL 7650  | 1.5  | 0.01  | 4        |

A more serious problem from a practical standpoint is the drift of the offset voltage with temperature. Since it is impossible to produce op-amps with absolutely identical characteristics on the two inputs, a zero input signal still results in a small output voltage. These offset voltages can always be trimmed to zero, and the crucial spec is then the drift with temperature. For example, if the temperature in the dome drops 20°C during the night, the no-signal output voltage above the dark current will change by a significant amount, especially on the higher gain steps. Of course, it is important to adjust the no-signal level above the actual zero level, with no PMT connected and the input shorted, and the zero offset should be adjusted to perhaps 5% above zero. Choosing an op-amp with an intrinsically small offset voltage drift is still important, because op-amps which have to be trimmed heavily in order to achieve zero offset then drift more than those which need to be trimmed only a little. Be careful to use a Cermet or other low temperature coefficient multiturn (15 to 20) potentiometer. Voltage offset drift can be a real annoyance.

Keep in mind that choosing an op-amp with good specs on input bias current and offset voltage drift is almost always a compromise between these two parameters. Another way out is to use a chopper-stabilized op-amp, which is discussed below.

There is one final note which pertains to "intermediate stages." Any analog output to a panel meter or other device should be buffered by a high-impedance op-amp to avoid any loading or other adverse effects on the actual signal.

### 5.4.4   Output Stages

The output stage of a PMT amplifier is a crucial part of the DC amplifier. A simple panel meter is inadequate, because of poor accuracy (about 2% for the expensive ones), poor resolution, and no capability to integrate or average the signal. The strip chart recorder was the standard output device during the 1950's and the fluctuating signal was averaged by eye using a clear plastic ruler. Fortunately, those days are gone. Heathkit offers a strip chart recorder in kit form for $350, but that seems like an expensive, outmoded, and inconvenient way to go. Several less expensive and better methods are now available. With the use of microcomputers, it is essential to provide the data in digital form in order to be read by the computer.

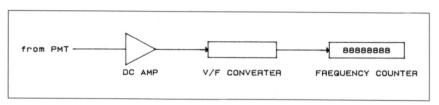

**Figure 5–8.** *Basic layout of DC amplifier and digital output device, using a* V/F *converter and frequency counter.*

The simplest digital output stage is shown in Fig. 5–8. The analog signal from the DC amplifier is converted to a frequency which is proportional to the signal. This frequency is then easily read by a frequency counter with a 10–second gate time which integrates or averages the signal over 10 seconds. Accurate frequency counters can now be built at home at a cost in the $50 to $75 range (for 8 digits), thanks partly to integrated circuits made for that purpose (e.g., the Intersil ICM 7216). Heathkit offers a counter at reasonable cost ($130), but it has only 0.1 and 1 second gate times. The gate time may be modified to 10 seconds (DuPuy, 1983a). Although the Heathkit counter is inexpensive and offers digital output in the sense that discrete numbers appear on the readout, there is no provision for interfacing to a microcomputer. Some (more expensive) frequency counters are available with BCD outputs that can be read by a parallel port. Another approach would be to have the microcomputer detect and count (via software) the frequency output of the V/F converter. The software for this would undoubtedly have to be assembly language or something like FORTH, in order to be fast enough. Of course, this would completely tie up the computer during integrations unless a "counter board" for the computer were purchased or built.

There is another aspect to the output device that is very important for a convenient DC amplifier: the dynamic range per gain step. The older DC amplifiers employed a dynamic range per gain step of 3×. For a total dynamic range of 10ᵐ0, it was necessary to calibrate 5 or 6 gain steps. That was necessary because of the limited dynamic range of the strip chart recorders or V/F converters. Not only the dynamic range of the input stage is involved, but one must consider the accuracy of the output stage over a large dynamic range. The latter is generally the limitation.

Let's briefly examine the limitation imposed by a dynamic range of a V/F converter. Suppose the linearity of the V/F converter is 0.1%, i.e., the output frequency can be relied on to be proportional to the input voltage to the third decimal place, and no more. That means that, for a 0ᵐ01 (1%) measurement precision, a dynamic range of 10 is permitted. As discussed below, good quality V/F converters typically have a non-linearity in the 0.01% to 0.05% range, which would permit gain steps of 100 or 20 respectively, for 1% precision throughout that range. Note that the offset error in the V/F converter is unimportant as long as the offset is stable and is in the same direction as the signal, since all photometric measurements are differences, even for all-sky photometry. With a dynamic range of 100, only one gain step ratio is required to be calibrated, and that is real progress!

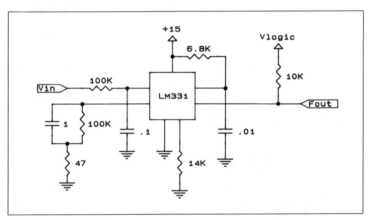

**Figure 5–9.** *Basic stand-alone* V/F *circuit for the LM 331. Adapted from the* National Linear Handbook. *Input voltage range is 0 to +10 volts, for output frequency range of 10 kHz full scale.*

Now for specific V/F circuits. Table 5–2 shows the non-linearity versus cost for three V/F converters. The AD 537 V/F converter has been available for several years and is an excellent performer (DuPuy, 1983b; Oliver, 1982b). The AD 537 is one of the simplest V/F converters to use, with only five external components required (three resistors, two capacitors) for positive voltage input. Note that several versions of the Analog Devices V/F converters are available. The AD 537K offers 0.03% (typical) non-linearity, and has better calibration specs (unimportant for our application) and temperature drift specs, but it costs $35 and requires a different socket than the J version. Two new and promising V/F converters are recently available. For sheer low cost, the National LM 331 is unbeatable! The 0.03% specification is for the simple stand-alone circuit as shown in Fig. 5–9, and their "precision" circuit should decrease that to 0.01%. In order to achieve precision at this level, care must be taken in several areas. For example, a stable low-temperature coefficient timing capacitor must be used; polystyrene is recommended. Thin film resistors are the best bet and DigiKey is a good source for many of these components. A "guard ring" on the input may be helpful: a pc board ring of copper at ground potential around the input terminal pins to reduce leakage currents. A new V/F converter from Analog Devices is the AD VFC32, and its specs and price suggest its use instead of the old favorite, the AD 537. See Table 5–2.

| TABLE 5–2 Comparison of V/F Converters | | | |
|---|---|---|---|
| Name | Linearity (at 10 kHz) | Cost ($) | Available |
| LM 331 | 0.03% | 4 | Jameco |
| AD 537JH | 0.1% | 8 | Analog Devices |
| AD VFC32KN | 0.01% | 9 | Analog Devices |
| AD 650KN | 0.002% | 15 | Analog Devices |

The newest and most promising high precision V/F converter is the AD 650. The non-linearity is typically 0.002% at 10 kHz! That suggests that one gain step could cover a range of 500 ( $6^m7$ ), with a precision of 1%. That means only one gain step ratio to calibrate for a dynamic range of $13^m4$. The circuit diagram for a "simplest application" using the AD 650 is shown in Fig. 5–10. For the best possible results, the digital and analog grounds should be separated, as shown in the spec sheet. Note that all V/F converters require a different circuit (same chip) if the input voltage is negative. The AD 650 probably should be utilized only after you have gained some experience with simpler V/F converters, since it may be tricky to realize the extra precision offered by this device. It is also worth noting that most "V/F converters" can be used as F/V converters, with simple circuit changes. That is very useful if you need to transmit an analog signal over a "long" wire, e.g., from the dome to the warm room, or even across the dome; simply convert the analog signal to a frequency, transmit the (now) digital signal, and reconvert it to an analog voltage. It is, of course, unwise to attempt to transmit a precision analog signal over a long wire. That is why it is risky to attempt to use a DC amplifier on the telescope, and then transmit that precision DC signal across the dome to the warm room or to the house. Any RF interference will be interpreted as part of the signal and, of course, this problem is worse for wide gain steps (e.g., 30×). There are safe methods to transmit this signal over a "long" wire, and converting to a digital signal is probably the most popular. The noise rejection properties of digital signals means that most interference has no effect on the signal. The V/F scheme works well here.

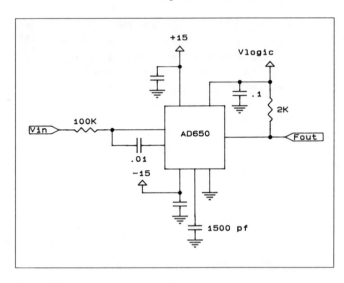

**Figure 5–10.** *Basic* V/F *circuit for the AD 650. Input range from 0 to +10 volts, for output frequency range of 10 kHz full scale.*

There is another, totally different output scheme which deserves mention. This one avoids buying a frequency counter but makes digital data available for the computer to read, all for less money than the old strip chart recorder! Instead of a V/F converter, the output from the input stage op-amp is routed to an integrator circuit with a 10–second precision timer. The integrator sums the signal over 10 seconds, thereby averaging it. This is an important step in normal (non-high speed) photometry. At the end of the ten-second interval, a sample-and-hold amplifier (SHA) holds the signal so it can be read by a digital panel voltmeter (DPV). In addition to yielding a low-cost digital readout, the DPV is available with binary coded decimal (BCD) output that can be read directly by a parallel computer port. The price of a suitable DPV varies considerably, depending on features and number of digits. Newport Electronics offers a 3 1/2 digit (e.g., 1.999 volts full scale) DPV with BCD outputs for $120. The disadvantage of this type of meter, compared to a V/F converter and 8–digit frequency counter, is the number of digits. With a resolution of only 1 in 1999, the gain steps will be limited to 20× (with a measurement precision of 1%) and, with some margin desirable, we are left with gain steps of perhaps 10×. DPV's with full scale readings of 3.999 volts are available from Newport for $220, and 4.999 volts for $250, but a 3.9999 volts DPV is out of sight at $500. A possible circuit for the DPV approach is shown in Figs. 5–11 and 5–12 but I have tested this one only under laboratory conditions, not at the telescope.

Note that the main incentive for using very high precision V/F converters is to avoid using and calibrating several gain steps. In practice, you can achieve observations with the same high precision with a lower precision (and less tricky) V/F converter, if you simply use a couple more gain steps. Even so, one can routinely depend on gain steps with a factor of 20 with the LM 331, thus insuring

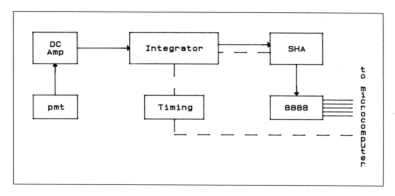

**Figure 5–11.** *Block diagram of integrator, sample and hold amplifier, and digital panel voltmeter as the digital output device.*

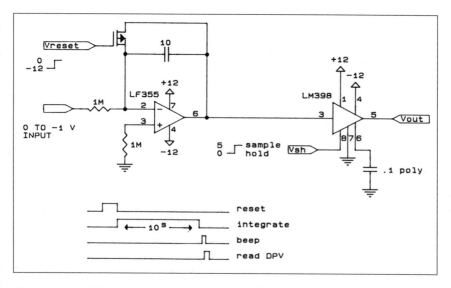

**Figure 5–12.** *Schematic diagram of experimental circuit for integrator, SHA, and DPV.*

a considerable margin of "safety" in the area of precision. This means that variable and comparison stars can generally be observed on the same gain step, thereby saving considerable effort and risk of recording the wrong gain step while observing and also offering higher precision.

The question of computer-controlled gain steps in a DC amplifier has been raised, and this is certainly an important question if completely remote operation from a warm room is anticipated using a DC amplifier. A satisfactory solution is not easy, because the stable gain of a DC amplifier depends critically on the stability of the feedback resistor. One simple and possibly satisfactory scheme is shown in Fig. 5–13, where an analog switch is used to select the appropriate feedback resistor. The analog switch chosen for experimenting was a CD 4016;

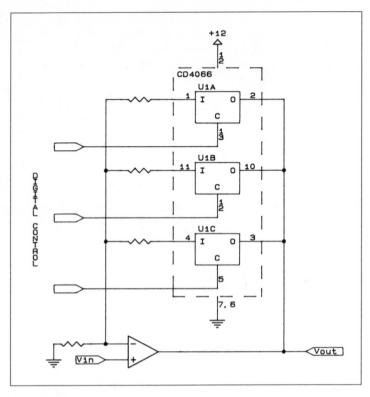

**Figure 5–13.** *Diagram of experimental circuit for computer control of gain steps in a DC amplifier.*

a CD 4066 ($0.39 from Jameco) is plug-in compatible and should be a slightly better choice because of "ON" resistance. Most analog switches of this type are characterized by a very high "OFF" resistance and a small but non-negligible "ON" resistance. The ON resistance is of concern here, since it is in series with the feedback resistor: how stable is this resistance? The spec sheets for the CD 4016 and CD 4066 suggest ON resistances of around 300 and 120 ohms, respectively. However, the ON resistance depends on the supply voltage. A simple test circuit on the breadboard indicated about 100 ohms for the CD 4016 with a +5 volt supply and around 20 ohms with a +12 volt supply, as calculated from an op-amp output. It is for this reason that +12 volts is indicated as the supply voltage on pin 14. This complicates the digital switching slightly (e.g., pins 5, 6, 12, 13), since the switch closes when a control signal equal to the supply voltage is applied. Assuming that most applications will use TTL or Schottky TTL logic, the simplest solution is to use open collector TTL drivers, e.g., a 7406. A TTL logic signal can then provide +12 volts to control the switch. Of course, it would be wise to make the ON resistance of the analog switch as small a contribution as possible to the total feedback resistance. Two cautionary comments: with the substrate (pin 7) grounded, a negative voltage must not be applied to the switch and, as with all CMOS chips, all unused inputs *must* go somewhere, e.g., be grounded. Although it is clear that the circuit shown in Fig. 5–13 works nicely for average signal levels of a few volts, the use of the analog switch with very large gains for nanoampere signals should be viewed as experimental. For example, what is the noise contribution of the analog switch in this application? If you are interested in this approach, I recommend careful testing with a constant current source.

After everything is considered, I recommend the general scheme shown in Fig. 5–8 (v/f converter and digital integration of signal) as the overall best approach with a DC amplifier.

### 5.4.5 Construction Techniques

A very helpful list of construction hints has been published by Oliver (1982a), and the comments in this section will build on these hints.

First, it is highly recommended to use a printed circuit (pc) board. A high quality fiberglass board (green) should be used. You will save considerable time and effort to obtain someone else's pc board layout. If you do not have access to materials to duplicate the board photographically, a 1:1 xerox copy of the layout may be used to locate the socket holes and other pads, using a small nail or punch. Then an etch resist pen can be used to draw the circuit onto the copper. Clean the copper first very carefully with a soap steel wool pad. Warm the etchant solution in a glass tray (do not use aluminum!) in a pan of heated water, and the etching should take about 20 minutes. The board can then be drilled with a #60 drill.

Be very sure that the DC supply voltages are "clean." The three-terminal voltage regulators are unbeatable (e.g., a 7812 and a 7912 for + and −12 volts, respectively), but use a quality bypass capacitor on the input and output, mounted very close to the regulator chips. It will help to mount the power supply components inside a separate aluminum enclosure or at least mount the transformer outside the enclosure containing the amplifier. If you intend to leave the amplifier in the dome, it is wise to maintain 110 VAC on the transformer, with a switch in the DC lines, to keep the instrument warm and moisture out. A small power resistor mounted inside the amplifier enclosure will help keep it dry inside; use a separate transformer for this to help avoid ground loop problems. Oliver recommends sealing the enclosure against moisture and leading wires in and out of the box with sealed grommets. Another approach is to mount a handle on the enclosure and store the amplifier inside the house where it is dry.

Gain step resistors should be mounted directly on the switch, with only two wires back to the pc board. The switch should be the ceramic type to avoid leakage currents. If very high value resistors are used (anything larger than a few megohms), avoid handling the resistors with your fingers, since finger oils will conduct leakage currents. A quality input connector is also required: BNC is okay, but a TNC (essentially a threaded BNC) is better.

If a normal op-amp is used (e.g., an AD 515), a multiturn potentiometer should be panel-mounted and connected to the op-amp offset adjust terminals. Be sure to use a low-temperature-coefficient potentiometer, such as Cermet. If the chopper-stabilized op-amp described below is used, an offset mounted on a pc board is adequate, since it is unlikely ever to need re-adjustment.

Most FET op-amps are uncomfortable with an input signal applied before power is applied to the op-amp; always turn the amplifier on and then plug in the signal cable. In operation, always begin with a lower (i.e., less sensitive) gain step and then increase the gain until the appropriate gain is reached. It is important that the output from the first stage be slightly above zero, so that the input to the V/F converter does not go negative under no-signal conditions (or conversely, if you wire your V/F converter for negative voltage). This is especially important if you are using a non-chopper-stabilized op-amp which could drift below zero under temperature changes. It is not at all necessary to try to adjust the amplifier exactly to zero; in fact, it is a risky practice.

### 5.4.6   A Chopper-Stabilized Amplifier

As an example illustrating some of the ideas above, an amplifier is described briefly; for more details, see DuPuy (1983b). The op-amp chosen for the input stage is the ICL 7650 chopper-stabilized op-amp. The chopper-stabilized aspect of this op-amp reduces the offset voltage drift with temperature to 0.01 $\mu$V/°C. See Table 5-1. The input bias current on the ICL 7650 is typically 1.5 pA, and that decreases with decreasing temperature.

A block diagram of the entire amplifier is shown in Fig. 5-14, and the complete schematic is shown in Fig. 5-15. The PMT current ranges from $10^{-6}$ to $10^{-10}$ amperes, and the ICL 7650 converts that to 0 to +4 volts (maximum output voltage on this device is +4.8 volts with +/−5 volts applied). The intermediate stage serves two purposes: (a) it serves as a 2.5× voltage amplifier to give a full scale output of +10 volts and (b) it serves as a zero offset adjustment, since the ICL 7650 does not offer a provision for this.

The ICL 7650 is used like any op-amp, except for two high-quality capacitors between pins 1 and 2 to pin 8; these set the chopping frequency. Pin numbers are shown for the 14-pin package. ICL7650CPD offers specs for 0°C to +70°C; ICL7650IJD specs cover −20°C to +85°C; and an 8-pin TO–99 can is also available.

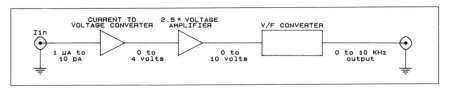

**Figure 5–14.** *Block diagram of chopper-stabilized amplifier, showing input current, intermediate voltages, and output frequency.*

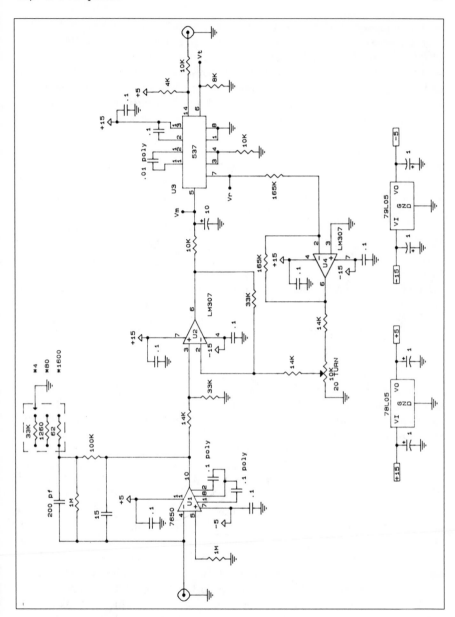

**Figure 5–15.** Schematic diagram of ICL 7650 chopper-stabilized op-amp DC amplifier (reprinted from DuPuy, 1983b).

The gain steps have been chosen with +10 volts full scale output on U2. Each gain step is approximately a factor of 20×, with full scale ranges of 1000–50 nA, 50–2.5 nA, and 2.5–0.125 nA. These cover the usual photocurrent ranges. For even brighter stars, an additional and less sensitive gain step is easily added by switching the gain switch to an open position (4× less sensitive). This circuit was designed with fairly slow response in mind: 0.5 milliseconds for the least sensitive gain step shown, and decreasing by a factor of 20× for each of the next two gain steps.

The V/F converter has been wired for positive voltages, and it is important under no-signal conditions that the near-zero output signal still be slightly positive. Hence, it was necessary to provide an external offset voltage adjustment. The AD 537 has a precision voltage reference on pin 7, which must be buffered, and U3 both inverts to −1.00 volts and buffers the reference voltage. U2 is wired as a differential amplifier, so that its output ranges from a positive near-zero voltage to +10 volts full scale. The offset voltage drift on the ICL 7650 is so stable that, once the 10K potentiometer is set, it is not likely to need adjustment again. Hence, this may be mounted on the pc board.

A discussion of V/F converter stages has been given above. The output frequency for the circuit shown in Fig. 5–12 is $f_{out} = V_{in}/(10R_tC_t)$, where $V_{in}$ is the voltage input to the V/F converter, and $R_t$ and $C_t$ are the RC timing components on pins 3, 4 and 11, 12. The frequency output is related to the PMT current by

$$f_{out} = 2.5 \times i_{pmt} \times 10^5(1 + 100,000/R_g)R_t^{-1}C_t^{-1}, \qquad (5.4.1)$$

where $R_g$ corresponds to $R_2$ in Fig. 5–7 (the gain multiplier resistor). $R_t$ is the 10K resistor on pins 3 and 4 of U4, which converts a 10-volt signal to a 1-milliampere input current for the V/F converter. No attempt has been made here to incorporate provisions to trim the gain and offset of the V/F converter to zero, since exact values are unimportant for this application (i.e., we don't care if 10.000 volts does not correspond exactly to 10.000 kHz). The output (pin 14) has a pullup resistor to 5 volts for proper interface to TTL circuits. A pc board layout is available from this author.

Three points have been marked on the diagram as $V_m$, $V_r$, and $V_t$. At these points a voltage signal is available for three options. $V_m$ denotes the appropriate place to obtain an analog signal to drive an analog panel meter. $V_r$ (pin 7, U4) denotes the 1.00 precision voltage reference on this particular V/F chip; it is being used in this circuit to provide a stable offset voltage. $V_t$ is an absolute temperature reference voltage, reading 298 millivolts at 25°C (298° Kelvin), with a change of 1 mV per °C. This will measure the chip temperature, which will likely be a few degrees above ambient dome temperature. All three of these voltage signals *should* be buffered with voltage followers and the $V_r$ signal *must* be buffered to avoid disturbing the V/F converter.

On any original design like this circuit, it is important to be sure the circuit performs as well as hoped! The main performance parameters of concern are the overall linearity and the effects of input bias current. For example, are the gain steps ( $3^m25$) so large as to compromise precision? Or is the input bias current large enough to cause problems for faint stars?

A Keithley Model 261 Picoampere Source was used to supply stable, known input currents to mimic a PMT. Measurements were taken over a range from $10^{-10}$ to $10^{-6}$ amperes, using three gain steps. The gain steps were carefully calibrated. A linear least squares solution was fitted with data points down to $10^{-9}$ amperes, and the logarithmic residuals were plotted versus the logarithm of the input current. The results are shown in Fig. 5–16, and there are no significant deviations from linearity, aside from about 1 pA input bias current which became evident when measuring 100 pA input current. The stability of the setup on the work bench was excellent. Hence, we conclude that the overall linearity is quite satisfactory, even with the large gain steps. Accurate measurements were obtained down to a level of 1/30 of full scale input, indicating that 20× gain steps is conservative. The input bias current around 1 pA is the only specification I would like to see improved on this amplifier, but it is small enough for most applications.

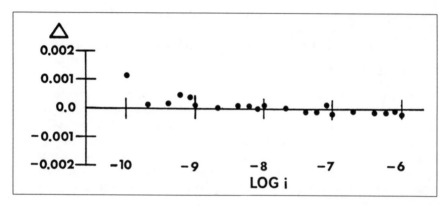

**Figure 5–16.** *Logarithmic residuals from a linear least squares fit to input current vs. output frequency, obtained with the ICL 7650 amplifier (reprinted from DuPuy, 1983b).*

## 5.5 Pulse Counting

### 5.5.1 Introduction

Direct pulse counting offers many advantages for those willing to observe the necessary precautions involved in handling high frequency, low amplitude signals. Under proper conditions, the best possible signal-to-noise ratio can be obtained with pulse counting. Problems in the calibration involved with DC amplifiers which use scales are greatly reduced in pulse counting. Sensitivity to changes in the high voltage supplied to the photomultiplier and to temperature effects can be reduced.

Under otherwise perfect conditions, the ultimate accuracy of photometry is limited by the random arrival of photons, which essentially obey Poisson statistics. The standard error of a single observation is inversely related to the square root of the count. Thus, a count of 10,000 is necessarily uncertain by $\sqrt{10,000} = 100$, which is 1% of 10,000, while a current of 1,000,000 is uncertain by at least $\sqrt{1,000,000} = 1000$, which is only 0.1% of 1,000,000.

At high count rates the recorded count rate $n$ underestimates the true count rate $N$ because some pulses are too close to each other to be resolved. Assume that every time the amplifier/discriminator detects a pulse it must wait the time $\delta$ before it can detect another pulse. If the rate at which pulses are being produced is $N$ counts per second, the number of pulses that will arrive during this dead time is $\delta N$. So, for each photon detected, $\delta N$ photons are missed and the rate at which photons are counted is

$$n = N(1 - \delta N). \tag{5.5.1}$$

This gives us three approximation formulas which may be used to correct for this saturation effect:

$$N = \frac{n}{1 - \delta N} \approx \frac{n}{1 - \delta n} \approx n(1 + \delta n). \tag{5.5.2}$$

Fernie (1976) discusses higher-order approximations to the dead-time correction.

### 5.5.2   An Integrated Circuit Pulse Amplifier and Discriminator [3]

### Introduction

A high-quality, low-cost pulse amplifier has been built for photon counting applications, with a single integrated circuit designed for this purpose. This section describes the specifications of the chip, details of the circuit, and notes on construction techniques. The only adjustment required is to set the threshold level. Commercially available pulse amplifiers for photon counting are generally somewhat expensive for small observatories, or for experimentation with pulse counting techniques. The cost of this unit is $50 to $100, depending on packaging.

### Features and specifications

The integrated circuit chosen for the pulse amplifier was the LeCroy MVL 100. This is a 16–pin DIP integrated circuit (IC) with a fixed gain of 100, an on-chip discriminator, and a bandwidth greater than 100 MHz. A few key specifications of the IC are tabulated in Table 5–3. As may be seen, this chip is capable of high quality pulse counting, for all but the most critical applications. LeCroy Research Systems offers other pulse amplifier IC's, some with more nearly "state-of-the-art" specifications; the choice of the MVL 100 for this application was based on the cost of the IC ($17) and a built-in on-chip discriminator.

The internal functional structure of the IC is shown in Fig. 5–17 (adapted from the LeCroy specification sheet). An analog output, with a gain of 10, is available for monitoring the signal output on an analog meter, if desired. The fixed gain of 100 is a limitation, of course, since some low-output photomultipliers (PMT's) require more gain than this. But for many PMT's in astronomical applications, a gain of 100 is appropriate. For example, this amplifier has been tested with EMI 6256 and EMI 6094 PMT's. (Another way of discussing sensitivity is to specify the threshold level; see discussion below.) As shown in Fig. 5–17, both negative and positive pulse outputs are available.The outputs are standard ECL levels (–0.8 to

---

[3]Written by David L. DuPuy (1981) and reprinted with permission from *P.A.S.P.*

–1.6 volts). These outputs may be offset to standard NIM levels. The capacitor-coupled output shown on the diagram was used to drive a Hewlett–Packard 5302A or a Fluke 1950A frequency counter directly, and this 0.8 volt pulse was adequate for reliable frequency counting in the dome.

| TABLE 5–3 | |
|---|---|
| KEY SPECIFICATIONS OF THE LeCROY MVL 100 IC | |
| Gain (fixed) | 100 |
| Bandwidth | >100 MHz |
| Output pulse width | 10 nsec–1000 nsec |
| Threshold voltage | 0.2–3.2 mV (adjustable) |
| Threshold stability | <0.2%°C |
| Output pulse level | standard ECL level (–0.8 to –1.6V) |
| Pulse rise time | 2.5 nsec |
| Pulse pair resolutions | 1.2× pulse width + 20 nsec |

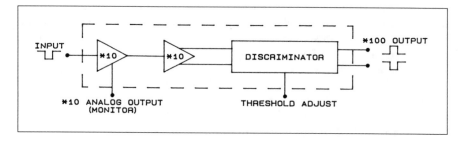

**Figure 5–17.** *A block diagram of the functional structure of the LeCroy MVL 100 integrated circuit, based on the LeCroy specification sheet.*

The schematic diagram of the complete pulse amplifier is shown in Fig. 5–18. The voltages required by the IC are +5 and –5 volts. We used three-terminal voltage regulators to provide these voltages. The threshold level of the discriminator is set by adjusting the voltage on pin 7 between –1.5 and –17 volts. We chose to supply this control voltage with a –12 volt IC regulator (on the assumption that a range of 0 to –12 volts was adequate) and a voltage divider, as shown in the diagram. A cermet 5 K ohms, 50 ppm/°C potentiometer was used.

## Construction and use

The circuit was built on an 8–cm square double-sided printed circuit (pc) board. The use of point-to-point wiring is not recommended. Several precautions are advisable (as with any wide-band amplifier) to avoid undesired instabilities and resulting oscillation. The purpose of the double-sided pc board is to provide a continuous ground plane on the component side of the board, with the printed circuit on the bottom side of the board. The ground plane was strapped and soldered at several points around the board to the etched ground strip surrounding the printed circuit on the bottom side. The printed circuit was drawn on the bottom side of the board, with an acid-resistant felt-tipped pen and with appropriate spacings for components. Crowding of components should be avoided to prevent unwanted coupling between components. Holes drilled through the board for mounting the socket and other components should be countersunk slightly to avoid shorts to the ground plane. A low profile IC socket should be used. The power supply was built on a separate pc board, along with a pulse generator for testing the amplifier (diagram not shown).

The power supply provides ±20 volts unregulated, with three-terminal IC regulators on the amplifier board. An AC power switch purposely was not provided in the transformer primary when the amplifier is left in the dome (although a moist, salty atmosphere may be a problem only for those living on the coast). Because it is important to provide sufficient decoupling of the power supply, separate low-pass filters are incorporated in each of the three voltage supply lines on the amplifier board. The bypass capacitors should be mounted very close to the IC regulators. The pulse amplifier IC runs quite warm during normal operation (almost too warm to touch), and the 0.1 ampere versions of the three-terminal regulators should not be substituted for the ±5 volt supplies. A shield between the power transformer/power supply and the pulse amplifier was not employed here, but it might be a worthwhile precaution. Before inserting the MVL 100 into its socket, it is worthwhile to check the output of the three IC regulators both for proper voltage level and for stability, with a resistive load and an oscilloscope.

The pulse amplifier was enclosed in a small steel box (2 by 4 by 8–inch Bud "Converta-Box"; $8) for protection from RF interference, with BNC input and output connectors. The line cord is not shielded, of course, and the AC line should be bypassed to ground ($0.01\mu$F disk) inside the enclosure. The third wire line-cord ground lead should be connected to the steel enclosure.

For testing, it is advisable to supply test pulses to the input (about 10 mV amplitude) and observe the output pulse with an oscilloscope. An inexpensive ($10) pulse generator can be built for this purpose, if a pulse generator is not available. The width of the output pulses is set by components on pin 1 and, for the values shown, output pulses are approximately 0.5 nsec/pF; the 100 pF capacitor shown produces pulses about 50 nsec wide. Other components can be used to produce a range of output pulse widths from 10nsec to 1000 nsec, according to the manufacturer. The threshold voltage (i.e., the discriminator level) may be set in the vicinity of –8 volts for the typical situation but may be adjusted for lower threshold voltages if desired, down to 0.2 mV. The amplifier is small and light enough to be mounted directly on the photometer on the back of the telescope. A short coaxial cable (e.g., one meter or less) between the PMT and the amplifier is desirable. The output pulses will drive a 50 ohm cable to a frequency counter. It is important to terminate this cable at the frequency counter with a feedthrough 50 ohm terminator, to avoid reflections and spurious counts. If spurious counts are encountered, RF interference from a nearby motor (e.g., a fan in the cold box or the dome motor) may be responsible. This amplifier has been used for normal photometry on the telescope with no apparent problems.

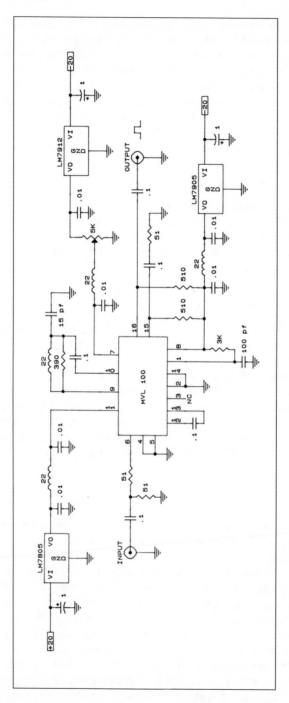

**Figure 5–18.** *Schematic diagram of the pulse amplifier described in the text in detail. All capacitors in units of pF unless otherwise specified. All resistors in ohms unless otherwise specified.*

Since this amplifier was built and tested, LeCroy has introduced an assembled and tested pulse amplifier which uses this IC (Model MVL 100 TB; $95). This unit includes LeCroy's pc board and components, input/output connections, and a small case; only a power supply and threshold voltage must be supplied.

It is a pleasure to acknowledge the hospitality of the Director and staff of Lowell Observatory, where this work was carried out. Dr. Art Hoag provided telescope time for testing this device and Mr. Ray Bertram assisted greatly in preparing instrumentation for the telescope. This work was supported by an Operating Grant from the Natural Sciences and Engineering Research Council of Canada.

### 5.5.3  Commercial Amplifier/Discriminators

The EMI Gencom AD–100 amplifier/discriminator is available for $200 (1985). It uses the LeCroy MVL–100 chip described in the previous section. Power requirements are ±5 VDC and the output is ECL. Somewhat more expensive amplifier/discriminators are also available commercially from EMI Gencom, Pacific Precision Instruments, and other manufacturers.

## 5.6  Data Recorders

Having discussed in the previous sections of this chapter the equipment required to produce a usable signal, we will now consider how the signal can be reco rded. In the area of data logging and control we encounter complexities and varieties of techniques greater than those encountered in any other chapter. The beginner is well-advised to start with the simpler approaches, and we say this not to be patronizing. Many of the observers who are most successful in terms of both the quality and quantity of their photometry use the simplest equipment for data logging. Thus it is appropriate that we start with the simplest possible system and, only after discussing it, move on in the following chapters to approaches of increased complexity.

### 5.6.1  Milliammeter

The simplest recording system consists of a milli- or microammeter as an indicator of the output of a DC amplifier. The meter must be read visually and the reading recorded manually. The meter needle will, of course, fluctuate constantly to some extent, because of scintillation, any variability in the sky transparency, and the statistical fluctuations inherent in the electrical current. Therefore, the observer must use his brain as an analog computer to determine the average reading over some interval of time, usually around 20 or 30 seconds. This is not at all a mark against meters, since the eye and brain are used in much the same way when reading a strip chart recording. It is not recommended that the observer lengthen the meter's time constant in an attempt to dampen the fluctuations completely, because a response too slow can prevent the alert observer from detecting danger signals, such as the star making a momentary excursion out of the diaphragm, a small cloud or patch of fog passing briefly in front of the star, etc. Although a simple meter is a useful indicating device, it is not without its drawbacks. Unless the meter is very large (and thus expensive), it is difficult to read it with precision.

The linearity and accuracy of cheaper meters is also questionable. Perhaps the most severe limitation is that the meter itself leaves no permanent record and so any question about a meter reading cannot be checked later.

## 5.6.2 Strip Chart Recorder

The strip chart recorder overcomes most of the deficiencies of the simple meter. A good recorder typically has a recording width of 10 inches, is quite linear, and can be read to an accuracy of 0.1%. Most importantly, it makes a permanent record of the observations. Moreover, it is quite easy to see, by a glance at the strip chart, when a star is drifting out of the diaphragm, a cloud is drifting by, etc. Also, by noting how closely comparison star deflections taken 10 or 20 minutes apart agree with each other, the observer can immediately gauge the photometric quality of the night as it progresses.

The same factors which cause the milliammeter needle to fluctuate cause the strip chart recorder's pen to oscillate and produce "grass" at the top of each deflection. Measuring each deflection is accomplished by simply drawing (literally or mentally) a horizontal line which passes as nearly through the middle of the grass as the eye can judge. As with the meter, the observer should not introduce a time constant which dampens the grass completely, for the same reasons.

Good strip charts need not be expensive. Many observers use the Heathkit recorder, which sells in kit form for around $425 (1982). Some diligent searching can turn up used recorders in good condition at very reasonable prices. A good speed for the paper advance is about 1 inch per minute. Much faster than this uses up lots of paper, while much slower does not allow sufficient space on which to record the star identification, gain settings, and other notations.

## 5.6.3 Counters

For systems which use an amplifier/discriminator as opposed to a DC amplifier, a digital pulse counter is used almost universally as the output indicator. Some systems with DC amplifiers also use counters after converting the DC signal into a pulse signal with a V/F converter. There are a number of advantages to digital counting systems. They, of course, have the capability of covering a very large range with very fine resolution. The inherent linearity of a counter is also very high, although the overall linearity and accuracy is most likely determined by what feeds the counter, not the counter itself. The counter must do its counting over some precise interval of time usually referred to as the integration time. This eliminates the need for determining the average of the fluctuations. Finally, the output from counters is in the numerical form required for data reduction and therefore avoids the not insignificant task of reading the strip chart.

Counters and their digital displays were originally large, expensive, and un-reliable devices employing many vacuum tubes. With the advent of LSI chips, counters are now both quite cheap and highly reliable. For these reasons their use is likely to become more frequent in the future. The Intersil ICM7226A is a good example of a modern 1–chip counter. It costs only $12 and can directly drive 8 digits of LED display. The only external components needed are the LED's, a MHz crystal, a couple of capacitors and resistors, a switch to select between 10,

1, 0.1, and 0.01 seconds integration time, and a switch to start the integration. A number of commercial counters are available starting at under $100.

### 5.6.4   STARLIGHT–1 Photon Counter[4]

We decided very early in the design of the STARLIGHT–1 that our stellar photometer would use photon counting techniques. When detecting very low light levels, photon counting offers optimum measurement accuracy and improved signal-to-noise ratios over DC measurements.

The presence of a photomultiplier output signal pulse of approximately 20 microamps will cause the amplifier/discriminator to output a 50–nanosecond differential ECL pulse into the counting electronics.

The photon counter has a 100–nanosecond pulse pair resolution which is commonly referred to as the dead time, $\delta$. As the mean pulse rate from the photomultiplier tube increases, a greater percentage of pulses will fail to be resolved by the preamp and counter electronics. The actual counting rate, $N$, can be approximated by Eq. (5.5.2), where $n$ is the number of counts per second displayed on the STARLIGHT–1. If $n$ is 1000 c.p.s., the percent error due to instrument dead time is only 0.01 percent, which is much smaller than the percent error due to standard Poisson counting statistics, However, at 100,000 c.p.s. the dead time error is 1.0 percent and the above correction should be applied to achieve maximum accuracy.

The STARLIGHT–1 uses an 8–digit LED display to show an output of up to $10^8$ counts. If further counts are received, an overflow indicator will light. (In practice, the overflow point would be reached only by making extremely long measurements of bright objects. Count rates of 1,000 to 10,000 counts per second would be typical for most applications.) The display has been designed to hold the previous reading while the next measurement is in progress. This allows 3 10-second measurements to be made and recorded within 30 seconds.

When using the digital readout, the user can choose one of two basic modes of operation—GATE and UNIT COUNT. In the GATE mode, the user can preselect counting time intervals of 10, 1, 0.1, or 0.01 seconds. At the end of the selected time interval, the accumulated counts are displayed. In the UNIT COUNT mode, counts are continuously updated to the LED display until the user signals the counter to stop. This mode is useful if the operator wants to use a single counting interval of longer than 10 seconds. The user can gate the counter manually through the START, STOP, and RESET push-button control switches which are mounted on the front panel and/or by using the remote control option.

The digital output has many desirable features over other modes of output. First, you do not have to invest in costly chart recorders, microcomputers, current metering equipment, etc. in order to determine accurate stellar magnitudes. The STARLIGHT–1 is a complete unit in itself; no accessories are required. Second, the average of your readings is an exact number, not an "eyeball" estimate of a moving needle or chart recorder output. The average background count is also an exact number to be subtracted from the variable, comparison,

---

[4]Written by Robert C. Wolpert

**Photograph 5–1.** *STARLIGHT–1 photon-counting circuitry is mounted on two photocircuit boards (one is coming out of the picture next to the front panel.)*

or check star counts. The background count is never lost in the "mud" at the bottom of an analog scale. Finally, you eliminate all potential sources of error which could be inherent in an additional piece of equipment. There is no need to change scales and set the zero point. The digital readout is probably the easiest output mode for an inexperienced person to work with and obtain accurate results.

We realized that there are many people who would prefer to use a chart recorder since it does have its own advantages of continuous data update and a permanent data recording. For these users, the STARLIGHT–1 provides a post-discriminator analog output signal which is completely independent of the digital counting system. In fact, both output modes can be used at the same time if it is desired! The analog circuitry provides a 1–volt full-scale output. Four ranges ($10^3, 10^4, 10^5$, and $10^6$ counts/second/volt) are provided with front panel switch-selectable control. A 10–turn front panel potentiometer permits full-scale zero suppression.

For users who prefer direct signal access, a buffered 50-nanosecond TTL output is available via a rear panel BNC connector. For maximum versatility, automation, and sophistication in data collection and reduction, the STARLIGHT–1 photon counter provides an external 25–pin I/O port for direct interface to a home or observatory microcomputer. In the UNIT COUNT mode, the computer will control the counting time interval; in the GATE mode, the STARLIGHT–1 timing circuitry (accurate to within 50ppm) will control the counting interval. In either mode, the computer can perform the necessary data reductions, corrections (dead time, extinction, transformation, etc.), and output format. The only limitation is the

user's programming ability. The counter unit is shown in Photograph 5–1 with
the top cover removed.

A new feature on the photon counter is the single/repeat switch. In the SIN-
GLE position the counter will behave exactly as described above. In the REPEAT
position, the counter will continually take measurements until the STOP button is
pressed. This provides greater convenience in recording data, especially when fast
timings are required. As an option, the STARLIGHT–1 counter can be supplied
with the 0–1600 volt power supply mounted inside the counter case. On the back
panel is located the MHV output connector, high voltage control pot, and a H.V.
monitor post. This feature allows the STARLIGHT–1 counter to be used with
any photometer head capable of single photoelectron resolution.

### 5.6.5   The Hopkins Phoenix Observatory Low Voltage Power Supply/Photon Counter [5]

### Introduction

The following circuits were designed for photon counting with the ECL (Emitter
Coupled Logic) output of the LeCroy MVL 100 Low–Level Amplifier/Comparator.
Provisions have been made also to allow TTL (Transistor Transistor Logic) inputs
from a DC amplifier and voltage-to-frequency converter (V/F). However, for the
cost of adding the V/F one can buy the MVL 100 and do photon counting. The
V/F approach adds a lot of convenience to photoelectric data gathering, as opposed
to a strip chart, but it still has a smaller dynamic range (requires scale changes on
the DC amp) and is much more susceptible to any variations in the high voltage
power supply to the PMT (photomultiplier tube). Photon counting, on the other
hand, can have up to eight decades of range with the eight digit counter and, since
pulses are counted instead of measured as an amplified current, any PMT high
voltage variations have considerably less effect.

### The Low Voltage Power Supply

The low voltage power supply shown in Fig. 5–19 uses a special transformer obtainable from Signal Transformer ($22–1982). The transformer allows either 115
VAC or 230 VAC 50/60Hz for input. The transformer's secondary is connected in
a dual complimentary configuration to obtain ±24V outputs. This configuration
minimizes the number of components and allows for a fairly compact power supply.
The 24V is used to allow for 3VPP voltage ripple, the rectifier voltage drop, the
voltage regulator voltage drop, and still provide enough margin for good regulation
in the event of a ±20% line voltage variation. By using two different regulators on
the –24V side, both –5V and –12V can be obtained. All three voltage regulators
are capable of supplying over 1 ampere. However, the –5V and –12V regulators
cannot exceed a combined current output of greater than 1 ampere due to other
circuit restraints. The current requirements for the counter and other circuitry

---

[5]Written by Jeffrey L. Hopkins

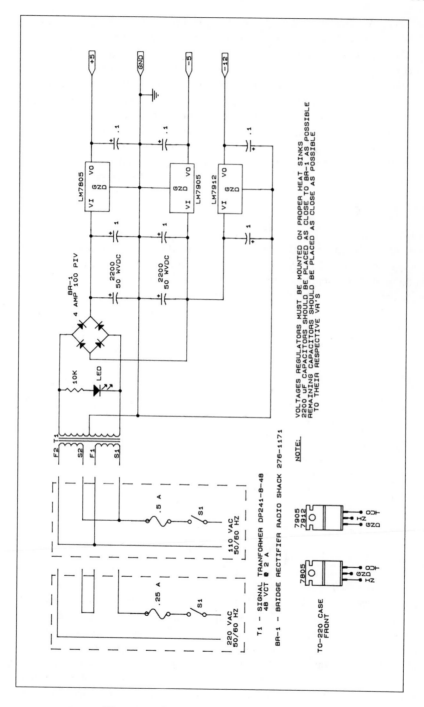

**Figure 5–19.** *Triple voltage power supply.*

are 500 mA (+5VDC), 100 mA (–5VDC), and 5 mA (–12VDC). As can be seen, there is quite sufficient current margin for the power supply.

Care must be taken when using the 7805 (+5V), 7905 (–5V), and 7912 (–12V) voltage regulators. The pin connections are different for the plus and the minus regulators. All three regulators should be mounted on heat sinks. The easiest way to accomplish this is to use mounting kits with insulators and mount the voltage regulators directly to the power-supply/photon-counter case.

Filter capacitors C1 and C2 should be mounted as close to the bridge rectifier BR–1 as possible. This is important in order to reduce the effect of any ripple on the voltage regulator ground terminal. The 1.0 $\mu$F and 0.1 $\mu$F capacitors should be mounted as close to their respective voltage regulators as possible.

### The Input Circuit

Although the Intersil ICM 7226A counter can accept TTL inputs directly, some protection should be used, because damage to the device will occur with excess input levels. To protect the ICM 7226A input and allow either TTL or ECL inputs to be used, the circuit shown in Fig. 5–20 was designed.

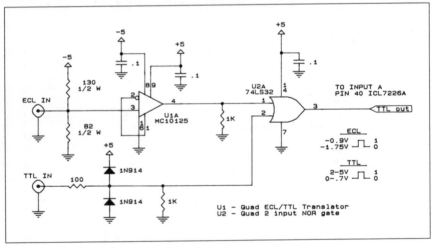

**Figure 5–20.** *Input circuit.*

For TTL inputs the two 1N914 diodes and 100 ohm resistor provide input protection. The MC10125 MECL–TO–TTL translator is manufactured by Motorola, Inc. It converts ECL (MECL) to TTL signal levels. There are four translators on the chip but only one is used for this application. The 130 and 82 ohm resistors should be placed as close to pin 3 as possible. This provides proper termination for the ECL input. The MC10125 incorporates differential inputs and Schottky TTL "totem pole" outputs. Differential inputs allow for use as an inverting/non-inverting translator or as a differential line receiver. A $V_{BB}$ reference voltage is available on pin 1 for use in single-ended input biasing as done here. The outputs go to a low logic level whenever the inputs are left floating. For non-inverting input, pin 1 ($V_{BB}$) is connected to pin 2 (inverting input) and the ECL input is

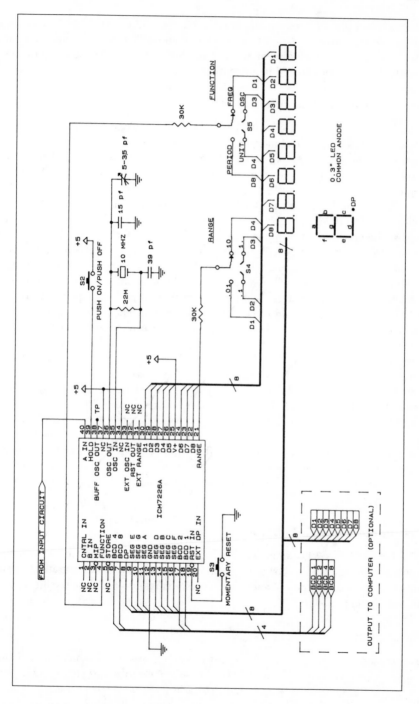

**Figure 5–21.** *10 MHz universal counter.*

connected to pin 3 (non-inverting input). The 74LS32 QUAD 2 INPUT OR GATE is used so that either ECL or TTL inputs can be used. When using the ECL input, the TTL input should be disconnected or at a low logic level. When using the TTL input, the ECL input should be disconnected.

## 10 MHz Universal Counter

The Intersil ICM 7226A ($32–1981) 10MHz Universal Counter System (IC1) provides a very convenient and inexpensive way to add an 8-digit counter to your photoelectric equipment. Intersil also has an evaluation kit available: 7226AEV/KIT (about $75). The kit comes complete with everything except a +5 VDC power supply and case and it can be built in about one hour.

The circuit shown in Fig. 5–22 is slightly modified for the photoelectric application. The ICM7216/7226 series counters are the first IC's to contain all the active circuitry needed to implement a Universal Counter on a single chip. The ICM 7226 integrates the oscillator circuitry, decode time base counter, 8-decade counter and latches, 7-segment decoder, 8-digit drivers, 8-segment drivers, control logic for multiplexing the display, and control logic for measurement and display updating. The counter input will function with an input frequency from DC to 10 MHz in FREQUENCY and UNIT COUNT modes and from DC to 2 MHz in other modes. If accurate absolute frequency is desired, a calibrated frequency counter should be connected to pin 38 (BUFF OSC OUT) and the 5–35 pF capacitor adjusted for exactly 10.000000 MHz on the calibrated counter. It should be noted that, when the ICM7226 is switched to OSC mode, it will display exactly 10 MHz, but this does not mean the oscillator is really on 10 MHz. The display is derived from the oscillator and will display 10 MHz no matter at what frequency the oscillator is. It should also be noted that for photoelectric work it is not necessary to have the oscillator exactly on 10 MHz. It is quite sufficient just to set the 5–35 pF capacitor to mid-position. By pressing the HOLD button, the display will remain until the HOLD button is pressed again. For typical V/F or photon counting, the FUNCTION switch should be in FREQUENCY mode and the RANGE switch in either 1 SEC or 10 SEC.

## Use of the LeCroy MVL–100

The LeCroy MVL–100a amplifier/discriminator is ideally suited for use with this counter. The schematic shown in Fig. 5–22 has worked well. Shown are the connections for both differential and single-ended configurations. Note the differences in terminating resistors.

## Conclusion

For under $200 a complete photon-counting system, less the photometer head and telescope, can be built. The unit is easy to build, versatile, and very reliable. For the computer enthusiast, a BCD and digit select are provided plus some control lines for the ICM7226. The amateur now has a very professional system available at a very modest cost. The completed unit is shown in Photograph 5–2.

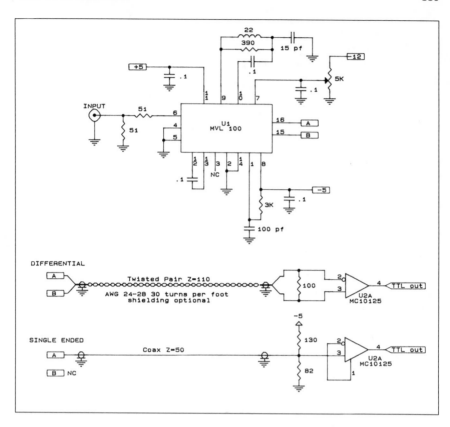

**Figure 5–22.** *Circuit for MVL-100. Capacitors are in µF unless marked otherwise.*

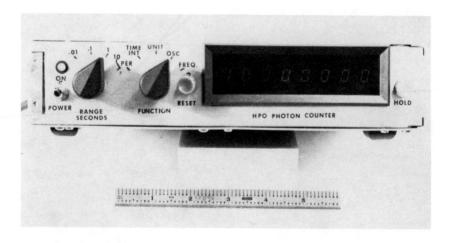

**Photograph 5–2.** *The Hopkins-Phoenix Observatory photon counter.*

# Chapter 6

# ADVANCED TECHNIQUES

## 6.1 Introduction

Most photometry at smaller observatories is done quite successfully without any significant automation except in the area of data reduction, where programmable calculators and small computers are being used with increasing frequency. In these cases, there is no direct connection between the photometer and the calculator or computer except that provided by the human eye which reads the meter, strip charts, or digital panel display and the fingers which key in the data.

For many smaller observatories the cost, trouble, and specialized skills required to interface a computer directly with a photometer generally do not make it worthwhile, although this situation is changing. Any increase in system complexity also is likely to add to the time the system is down for repair. For those with the necessary funds, skills, and patience, limited computerization might be considered after experience is gained in the operation of a conventional system. However, there is a danger that the time spent in such development could seriously jeopardize the observing program.

The goal of computerization in photoelectric photometry is to improve the efficiency with which the data are gathered and analyzed. Efficiency can be improved in three ways. First, the percentage of acceptable data points can be increased by reducing errors in instrument settings, data recording, and analysis. Second, the time taken to make an observation can be decreased by shortening the time taken to do the individual basic operations in a sequence, by reducing either the number of individual operations required or the number of telescope movements, or both. Finally, the time spent in data reduction and analysis can readily be reduced by computerization.

## 6.2 Guidelines for the Use of Computers

The more efficient approaches to photometry require automated control of a number of functions and, in some cases, require real-time mathematical analysis for adaptive control and for quality checks provided to the operator. While the control functions can be hardwired, Kibrick, Rickets, and Robinson (1979) point out that microcomputers are more reliable than equivalent hardwired logic. They are easily changed via software and they are now so cheap that a backup computer is possible. The use of a second, identical microcomputer at a different location for software development is also made possible by the low cost of these units; this has been reported by Honeycutt, Kephart, and Henden (1978) for the system at Goethe Link Observatory.

While the use of computers has many advantages, there are two major problems. First, as pointed out by Lasker (1972), the dome of an observatory is not environmentally conducive to the reliable operation of computers. Second, computers obviously must be programmed. While the solution to the first problem requires only a remote computer, away from the telescope in a warm room or in a warm box, the problem of programming is a serious one.

Based on their experience in developing the Lowell Observatory data system, Albrecht, Boyce, and Chastain (1971) strongly recommend development of the software in small modules, as does Lasker (1972). Robinson (1975) discusses the use of assembly, compiler, and interpretive languages in astronomical control. The limited memory capacity and speed of earlier computers often necessitated the use of special assembly languages peculiar to each type of computer At the same time, the low-cost of larger memories and faster machine speeds has made the use of higher level and more universal languages possible in real-time control tasks such as those in photometry. A fast, high-level language called FORTH has been developed by Moore and Rather (1973). Although developed for real-time astronomical control, it has been used in a number of other disciplines. The use of the highly interactive and easily programmed interpretive BASIC for real-time control has been reported by Titus (1979). While not nearly as fast as FORTH, interpretive BASIC is the most widely known and easiest to program language available. Where high speed beyond the capabilities of BASIC is required, small machine language subroutines can be included, or one can consider a compiled BASIC.

It is necessary, of course, to establish communication between the operator and the computer-controlled system. Lasker (1972) suggests that the number of switches and displays be minimized, that plain English (or French, etc.) be used, that the inputs be checked for correctness, and that an unconditional interrupt be provided for return to the main menu when all else fails. For the lone system operator on a high perch in a cold dark observatory, sole input to the system via a small hand-held keypad and sole output via a video monitor on a movable stand is probably ideal. Only fully remote operation from a warm room is likely to improve on this.

Finally, one must consider the selection of the computer itself. While for many years minicomputers, such as the PDP–8, were the most popular choice for real-time astronomical control, the use of microcomputers is increasing rapidly. In either case, another very important consideration, pointed out by Robinson (1975), is that only those computers which have sold in large quantities for several years are likely to be well supported by reliable peripherals and good software.

A detailed discussion of these and other considerations is contained in the books by Genet (1982) and by Trueblood and Genet (1985).

## 6.3 Approaches to Computerization

A number of different approaches to the goal of increased efficiency are possible. These will be illustrated by brief descriptions of the salient features of a number of pioneering systems.

The conventional system with a strip chart recorder is an example of a single-channel system without any automation. A low-cost but significant improvement on the conventional system has been made by Howard Louth. Data analysis in the Louth system is fully computerized on an HP–67 programmable calculator (later replaced with an Apple computer). The timing of the observations is automatic (fixed) and the photon count and time are printed automatically. These control features are implemented with a simple TTL circuit. The number of telescope movements per differential magnitude is reduced from 4 to 2 by the use of a manually operated offset of the diaphragm, which allows the background to be observed without shifting the telescope.

Davidson, Neff, and Enemark (1976) have taken the next logical step by automating the diaphragm offset, although this system does not print out the time automatically. The diaphragm movement in the photometer is controlled by a solenoid. Fernie (1975) achieved the same end by using a rotating mask which alternately uncovers two closely spaced diaphragms, one for the star and one for the background.

Instead of automating the offset to the background, the filters can be sequenced automatically. Sorvari (1975) developed a photometer in which the filters are sequenced automatically by a stepper motor which is controlled by simple TTL logic.

## 6.4 A Microcomputer-Based System for Photoelectric Photometry

### 6.4.1 Introduction

This section of this chapter describes a computerized data logging system developed by one of us (Genet) that was in operation for three years at the Fairborn Observatory starting in the late 1970's. The system used the then low-cost and readily available Radio Shack TRS–80 microcomputer. Based on the anticipated uses of the microcomputer, it was decided early on that the computer would need to be physically located in a warm room and operated remotely from the observatory when used for photometry. To allow programming in BASIC and to maintain maximum future flexibility, it was decided to interface the microcomputer to the outside world via a general purpose interface that made use of the recently available LSI (Large Scale Integration) programmable peripheral interface chips and also contained a real-time clock independent of the computer's clock. To maintain its generalized nature and flexibility, it was decided to channel all observer/computer communications via a remote hexidecimal keypad and video monitor.

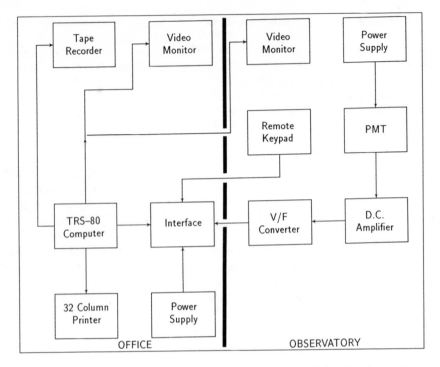

**Figure 6–1.**  *Block diagram of the Fairborn computerized data logging system.*

## 6.4.2   Equipment

A block diagram of the system is shown in Fig. 6–1. The equipment contained in the observatory is shown on the right side of the diagram, while that located in the office is shown on the left. The photometer head is permanently mounted on the 20–cm $f/15$ Cassegrain telescope. It contains a previewer with wide angle eyepiece, diaphragms of various sizes mounted on a wheel, a microscope for centering the star in the diaphragm, filters on a wheel for UBV and red leak, Fabry lens, and 1P21 photomultiplier.

The computer is a TRS–80 Model I, Level II. The computer is actually contained within the keyboard unit. The system was developed and operated without the expansion interface, disk drive, and modem; these were added just as an operating convenience and with an eye towards meeting anticipated future requirements. The interface contains only 6 chips. Three chips are 7404 hex inverters used for the clock oscillator and for generation of chip select signals. Two chips are Intel 8253 Programmable Interval Timers (PIT). Each of the 8253 chips contains three independent 16–bit counters that are directly programmable and readable by the computer. A special feature of these PIT chips is that they can be "read on the fly" without disturbing a count in progress. The final chip is an 8255 Programmable Peripheral Interface (PPI).

The 1 MHz oscillator is the basic local time reference. Two of the counters in one of the PIT chips are used to divide it down to 1 KHz and 1 Hz references, while the third one is used as a delay adjustable via software in one millisecond steps to achieve synchronization with the count of the seconds elapsed since clock initialization. Another PIT chip counts the pulses arriving from the V/F converter. The third PIT chip precisely times the integration (software setable). The 8255 PPI chip reads the keypad encoder, issues various discrete or checks status flags, and has sufficient remaining capability to control other functions.

### 6.4.3   Operation

The operation of the system will be described by going through an actual sequence. The computer is powered up simply by turning it on. The operating program is loaded off the disk and the JD (last four digits), month, and double date are entered. At this point, control is automatically transferred to the observatory, the video monitor and lights are turned off in the office, and the door is latched to preclude unauthorized access by junior members of the staff.

Upon arriving in the observatory the "Main Menu" is displayed on the remote monitor. The time cube is activated and set on a post outside the observatory to avoid the considerable interference from the computer that travels along the control lines. Option "1" from the menu is selected. For "slow" variables close synchronization of the local clock with WWV is not required and any three-digit delay may be keyed in. For somewhat faster variables the WWV "tick" and local clock "tick" may be synchronized by ear to within 1/10th second. More precise alignment, such as required for occultations, requires an oscilloscope to observe the WWV tick second crossover.

Once the WWV and local clock seconds ticks are aligned satisfactorily for the type of observation, the clock is set by asking for menu option "2". The computer asks for the hours and minutes (UT) of the next "at the tone." These are entered on the keypad and, when the tone is heard, full synchronization will be maintained if any key is pressed within a second. The time is then displayed in the upper right-hand corner of the screen. It may be checked at any later time by asking for menu option "7".

With the clock now set, menu option "3" is selected, the list of the current program stars is displayed, and a pair (variable and comparison) is chosen. The list of program stars is updated about once a month, adding new ones in the East and dropping old ones in the West.

A sequence of observations on the selected star pair (differential photometry) is then initiated by selecting menu option "4". The computer asks "number of colors?" While the computer can handle any number of colors, the remote video monitor can display only 3 or 4 conveniently. Assuming 1 color (V) is desired, a "1" is entered on the hand-held keypad. The computer then responds with the observer prompt which includes: identification of variable and comparison stars, whether the variable or comparison is to be observed, whether the star or the background is to be observed, and the color of the filter.

The observer (without computer assistance) must then find and center the indicated star (or sky offset) and select the indicated filter. At the beginning of any sequence the observer must also adjust the gains on the DC amplifier so that the brightest star through the "brightest" filter is near full scale. Once everything is set, pushing any key starts the precisely timed integration. As the pulses from the V/F converter are being counted, the counter is read "on the fly" and this reading is used to generate on the video monitor a display of light intensity versus time, similar in many respects to the display on a strip chart recorder. Such abnormalities during the integration as the star drifting out of the diaphragm, a star drifting into the diaphragm (during background), an unnoticed cloud, an accidentally kicked cable, etc., can be seen easily on the display. At the end of the integration the final count is displayed along with the time. As with all other displays, pushing any key advances the observer to the next step or returns him to the main menu as appropriate.

In this case, the next step is a display of all the light intensity counts and observation times in an array. The Comparison Star (CS) counts are all in a single column as are the Comparison Background (CB), Variable Star (VS), and Variable Background (VB) counts. This allows one to tell at a glance how the latest count compares with all previous counts in the sequence. It is immediately apparent if one is observing the wrong star, forgot to flip down the mirror during the background, etc. One notes mentally whether the reading is "out of line" or not. The computer then asks for a choice: (1) Terminate Sequence, (2) Repeat Observation, or (3) Data OK Continue. One can terminate at any time. If the just completed observation is questionable, it may be repeated. If everything is proper, pressing "3" brings up the prompt for the next observation. The combination of the video display, the count array, and the ability to repeat or terminate as well as continue, has proven to be very useful. This observer, at least, would have a difficult time going through an entire observing sequence late at night without making an error of some sort, so error detection and recovery procedures are an important prerequisite in a computerized system.

By means of prompts, displays, and commands, the sequence on a pair of stars is completed. The order is always CS CB VS VB CS CB VS VB ...... CS CB. The sequence always starts with a prompt for CS. If a repeat is asked for, the current prompt is repeated. Although the sequence may be terminated at any point, one usually ends a sequence with CS and CB, which are required for bracketing in differential photometry. Terminating a sequence automatically returns one to the main menu.

To retain the raw counts, times, and identifying information for the record, menu option "8" is asked for and the data are automatically printed out as shown below:

COMP/VARBL        47B00/44B00

| CS/UT | CS/CT | CB/UT | CB/CT | VS/UT | VS/CT | VB/UT | VB/CT |
|-------|-------|-------|-------|-------|-------|-------|-------|
| 31702 | 11609 | 31754 | 3234  | 31913 | 20474 | 32003 | 3256  |
| 32112 | 11205 | 32207 | 3296  | 32312 | 19348 | 32404 | 3292  |
| 32505 | 11153 | 32558 | 3321  | 32916 | 19781 | 33024 | 3269  |
| 33256 | 11297 | 33346 | 3282  | 33449 | 19741 | 35548 | 3284  |
| 33701 | 11451 | 33754 | 3253  | 0     | 0     | 0     | 0     |

These are actual data on a not-so-good night. The columns are Comparison Star Universal Time (CS/UT), Comparison Star Count (CS/CT), etc.

By asking for option "9", the raw differential magnitudes can then be calculated and both printed and displayed. The printout is as shown below. Note that the mean and the standard deviation of a single measure from that mean are given for the sequence. The standard deviation has proven useful as a gauge of the photometric quality of the night, on occasion resulting in observatory shutdown.

```
FAIRBORN OBSERVATORY PEP ANALYSIS - 1980
JUL DATE = 4394  MONTH = 6  DAY1 = 3 DAY2 = 4
COMP/VARBL - 47B00/44B00  FILTER = VISUAL


     UT       JD     D MAG


   31913    .63834   -.813
   32312    .64111   -.774
   32916    .64532   -.797
   33449    .64917   -.768


AVERAGE = -0.788
STD DEVIATION = +/-0.021
```

At this point, another pair of stars can be selected or the observatory can be shut down. Prior to shutdown, the clock is usually checked again against WWV as a precaution. It is not unusual to do the final reduction of the data on the same night as the observations are made. Finally, the computer printouts are stapled to the observatory log book and various remarks (observing conditions, equipment notes, etc.) are written in by hand.

## 6.4.4   Evaluation

It was found that this system did not significantly speed up the actual process of making the observations, but, by totally eliminating backlogs of unread strip charts and unreduced data, it increased the amount of time spent observing. It seems that a pile of unread strip charts was a real deterrent to observing, at least for this observer (Genet).

It has been almost a decade since development of the system described above was initiated, yet the approach taken today is usually similar. The greatest differences are that: (1) photometers are commercially available now that put out TTL pulses; (2) many microcomputers have built-in programmable counters and clock/calendar chips; and (3) microcomputers now cost a tenth of what they did and do much more.

The reader interested in developing a computerized data-logging system for photometry would do well to examine the back issues of the *IAPPP Communications, Advances In Photoelectric Photometry, Volumes 1 and 2,* and *Microcomputers In Astronomy I and II.* These contain many papers which give detailed instructions for interfacing commercially available photometers (such as the EMI Gencom STARLIGHT–1 and the Optec SSP–3) with popular personal computers (such as the Apple, the IBM–PC, and the Commodore 64).

## 6.5  Dual Channel Systems

### 6.5.1  Introduction

While single-channel systems have achieved a high degree of operational efficiency, they are limited at locations that have a high frequency of rapidly varying transparency. Dual-channel systems have been designed to circumvent this handicap. These systems tend to be complex and prone to various subtle systematic errors, however, and should not be considered until one has considerable experience with single-channel systems.

Fernie (1979) has described very clearly the advantage of a dual-channel system: "It is the night your binary is going into its once-a-century eclipse, the sun has set on a sky of the most incredibly clear emphrean blue, the equipment is stable to a zillionth of a magnitude, your fingers twitch in anticipation, and then, slowly out of the northwest, comes creeping like some inexorable slime in a science fiction movie, a great sea of high thin cirrus. What you need on these all-too-frequent occasions, primal screaming apart, is a second channel to monitor the sky transparency while the first channel whacks away at the binary." Apparently atmospheric conditions are less than ideal in Scotland also, as Reddish (1966) of the Royal Observatory in Edinburgh writes: "Changes in the sky transparency between successive observations are a dominant source of error in photoelectric photometry of stars." From Italy, DeBiase et al. (1978) state that "the atmosphere is the most important source of errors in ground-based astronomical photoelectric photometry."

By making simultaneous or nearly simultaneous observations of the variable and comparison stars, the dual-channel system is able to eliminate the effects of changes in sky transparency which affect both stars by virtually the same amount at virtually the same time. For large telescopes observing faint stars, the angular distance of a suitable comparison star is never very great. This has two important consequences. First, it is possible to observe both the variable and comparison stars at the same time without shifting the telescope, as they will both be within the field-of-view of a single telescope. Second, at such small angular separations, changes in atmospheric transparency affecting one star are bound to affect the other star almost equally and nearly simultaneously, as established by Bensammar (1978). These conditions are necessary for the optimum operation of any dual-channel system.

Small telescopes are necessarily limited to brighter stars and the angular distance to a suitable comparison star typically is much greater than it is for large telescopes. If this distance exceeds the field of view, two telescopes would be required for simultaneous observations of both stars. A single mount and observatory structure can, of course, be used for both telescopes.

It is not at all obvious that, at the 0.5– to 2.0–degree angular distances involved in differential photometry of bright stars, changes in atmospheric transparency will be sufficiently correlated for proper dual-channel operation. It is possible that the solar pulsations reported by Hill and Severny were pertinent to the question of the correlation of atmospheric attenuation. Research by Grec and Fossat (1979), Clarke (1978), and Fossat et al. (1977) suggests that the observed pulsation of the Sun was due not to an actual change in the size of the sun, but rather to the high correlation of changes in the earth's atmospheric transparency. Fernie (1979) reports good results with a two-telescope dual-channel system at angular separations of 2 degrees. Thus, it appears that variations in atmospheric extinction are sufficiently correlated, at the angular separations required by differential photometry of bright stars, to make the operation of small dual-telescope systems possible.

From the above discussion it can be seen that dual-channel systems can be segregated into two categories: one-telescope and two-telescope systems.

### 6.5.2   One-Telescope Dual-Channel Systems

A number of dual-channel systems employing a single light sensor (photomultiplier tube) have been developed. The advantage of using a single sensor is that the peculiar response of the sensor will be applied equally to both the variable and the comparison star, thus minimizing any effect on the differential measurement. The disadvantage of using a single sensor is that a somewhat complex electro-mechanical device must be made to switch the sensor back and forth rapidly between the variable and the comparison star. A system which does not only this switching but also the switching of seven colored filters has been developed by Burnet and Rufener (1979). The system uses rotating and oscillating wheels and mirrors under computer control to achieve this. The photometer head weighs 80 kg but is intended only for larger telescopes. A very interesting feature of this system is the use of a computer to calculate special statistics on-line from observational samples. This is used to warn the operator of possibly bad readings, such as might be caused by the star drifting out of the diaphragm center.

A relatively simple two-channel system using two sensors has been developed by Geyer and Hoffman (1975). Two separate photometers are mounted on the tailstock of a telescope. Provisions are made for rotating and translating the photometers so that almost any two stars in the telescope's field of view can be observed simultaneously. While the control of the filters and data logging of this system have not been automated, these functions have been fully automated on the system developed by DeBiase et al. (1978). The DeBiase system uses real-time computer analysis to optimize the observation times for each filter. As pointed out by DeBiase and Sedmak (1974), this is a case of "automatic adaptive control." A very similar system has been developed by Bernacca et al. (1978), although the emphasis is on high-speed observations and precision timing.

A one-sensor dual-channel photometer has been developed by Seeds (1970). The entire unit rotates to a line between the stars and the two channels are spaced equally from the center. The main channel always remains above the photomultiplier, while the second channel is at varying distances. Relay lenses on the second channel allow for this variable spacing. A moving prism is used to chop rapidly between the two channels.

To date, the most widely used dual-channel system is that of Nather (1973), there being a dozen systems in operation or under construction. Basically, this two-sensor system operates with the main channel similar to any one-channel system, and the second channel on an X, Y carriage operating from an offset guider box. The filter wheels are operated by stepper motors and the entire system is controlled by a minicomputer. While conceptually simple, the success of this photometer lies in the careful evolutionary refinement of both the hardware and computer software.

While a number of other one-telescope dual-channel systems have been developed, those discussed above are representative of what has been done so far.

### 6.5.3   Two-Telescope Dual-Channel Systems

To our knowledge only three dual-telescope photometric systems have been described in the literature. The first was in operation at Dunsink Observatory near Dublin, Ireland, over 35 years ago (Argue and Butler 1952). It was not in any way a computerized system; moreover, it has not been in operation for many years.

A dual 16–inch telescope equipped for photoelectric photometry has been reported by Reddish (1966). At the time of the report many of the operations had been computerized, including computerized movement of the telescope for the background measurements and changing of the colored filters. This telescope has recently been moved to St. Andrews.

Fernie (1979) has reported some preliminary results on the use of a temporarily configured system consisting of two totally separate telescopes at the David Dunlap Observatory. His results, made at angular separations up to two degrees and through thin cirrus clouds, are very encouraging for those photometrists who wish to use a dual-telescope photometric system to overcome the disadvantage of poor atmospheric conditions.

Between the first and second editions of this book, Fernie has proceeded to develop and operate a highly capable two-telescope system that is now routinely used to observe variable stars to very good effect. Other two-telescope systems are under development.

# Chapter 7

# AUTOMATIC PHOTOELECTRIC TELESCOPES

## 7.1 Introduction

Computerized data logging and analysis greatly relieves the tedium involved in photoelectric observations of variable stars. It still, of course, leaves the astronomer out with the telescope, moving the telescope to the stars to be observed, centering the stars, starting the integrations, and changing the filters. Quite frankly, many astronomers enjoy being out under the stars and using a telescope and photometer with a high degree of skill honed over years of making observations. Familiar groups of stars become old friends, and a peaceful evening's work allows one to unwind from the more hectic activities of the day.

This idyllic situation is the case for many photometrists around the world: amateurs who spend a couple of hours in the evening for perhaps 50 to 100 nights per year following a few favorite long-period variables; students at smaller colleges and universities who spend a few clear nights obtaining a light curve of an eclipsing binary; or the professor in the physics department of a modest-sized university who spends a week at a national observatory doing photometry on some stars of special interest followed on occasion over the years.

There is, however, a less idyllic situation. Here, one has a keen interest to gather data but, because of work and family commitments, there is either insufficient time to make the observations or, if the observations are made, there is a tendency to sleep on the job or slight both family and friends. When faced with a conflict between doing science and keeping the work and family situation intact, it is not surprising that a number of inventive minds have considered automation of photoelectric photometry of variable stars. The objective of such automation is simply to allow the astronomer to sleep while the data he wants are gathered by his trusty assistant—the automatic photoelectric telescope.

It would, however, be only fair to note that there is another, perhaps easier, solution than automation which allows the astronomer to sleep at night: get others to stay up at night and make the observations for you. This has had at least two successful forms. A large number of different observers spread about the country or world each can make a few observations for you on the same medium-or-long-period variables. With several dozen observers spread about in different climates, the coverage on a dozen variables can be quite good. Each observer need make observations for just a couple of hours in the evening for only a fraction of the nights. Everyone gets some sleep. This approach for doing science and sleeping has been developed to a fine art by one of us (Hall), as well as by a few others.

While the approach described above works well on medium-or-long-period variables, it does not work well on short-period variable stars, asteroids, etc., where all-night observing is required. Here, the second approach developed by Kenneth Zeigler and his associates at the Gila Observatory in Arizona should be considered. In this approach, the astronomer (Zeigler) supplies the telescope, equipment, and facilities. The astronomer also plans the observations and is the principle reducer of the data, writer of the papers, etc. However, the observations themselves—in this case of short period asteroids—are made by students from the high schools in the area of Globe, Arizona. No single student has to stay up all night—they take turns—nor do students have to observe every night—again they take turns. Collectively, however, they are able to keep the system operating almost every clear night, of which Arizona has many, and the Gila Observatory has become a major source of asteroid light curves and discoveries in this area of astronomy. A somewhat similar situation, but one developed as early as 1969, exists for various types of variable stars at the Auckland Observatory. Here W.S.G. Walker and Brian Marino have, over the years, assembled a fine team of observers.

For those astronomers more inclined towards computers and mechanical contrivances than to the wide-scale organization of observers, automatic photoelectric telescopes represent a viable option. However, unlike computerized data logging and analysis systems, which can and should be fairly simple, automatic photoelectric telescopes are fairly complex devices. One must not only log and analyze the data but must also control the photometer (switching the filters, etc.) and also control the telescope so that it can find and center in a very small diaphragm the specific stars desired (out of many millions). Moreover, this must all be done automatically so that the astronomer can sleep at night; otherwise the entire purpose for such automation is lost.

In actual practice, as we shall see, the true reasons for the development of automatic photoelectric telescopes have been almost as varied as the pioneers in this field. While it is true that some astronomers motivated by the desire to sleep at night have been involved in the field, engineers with a desire to develop automatic systems for the fun of it have had an important influence. In the very beginning it was the desire of two institutions to get into space astronomy that motivated APT development.

Also, as we shall see, the power and capabilities of automatic photoelectric telescopes were much greater than originally conceived. While it became true that astronomers could sleep at night, this ended up being of secondary importance compared to the new approaches to observational astronomy that full automation opened up. The power of an assemblage of fully automatic machines observing tirelessly from a prime photometric site is allowing entire classes of variable stars to be observed, and new programs involving the monitoring of hundreds or even thousands of stars for periods of years to be seriously contemplated.

## 7.2 A Short History of Automatic Photoelectric Telescopes

The history of automatic photoelectric telescopes (APT's) began in October 1957 when Sputnik was launched. Very soon NASA was formed and two organizations in particular began thinking about telescopes in space. These organizations were the University of Wisconsin and the newly formed Space Division of Kitt Peak National Observatory. They chose the development of automatic photoelectric telescopes as a means of getting into the space telescope business.

The Wisconsin University APT effort was headed by Arthur D. Code who, following in the footsteps of Joel Stebbins and Albert Whitford, was Director of the Washburn Observatory. Code thought that a good entry into space astronomy would be the development of a totally automatic telescope on the ground devoted to some fairly simple task. The simple task chosen for their APT was the observation of a series of bright standard stars each night to establish nightly extinction coefficients to be used in the reduction of the observations made by the larger telescopes at the Washburn Observatory.

The approach taken by Code and his associates was to obtain, as surplus equipment, a small automated (servo controlled) telescope, mount, and control system used for the alignment of guidance systems on Titan Intercontinental Ballistic Missiles. A photometer was added to this and an attempt, most unsuccessful, was made to get it all to work as an APT. In a successful revision of the system, the small refractor was replaced with an off-axis 8–inch prime-focus reflector/photometer (exactly as used in the OAO space telescopes) and the analog control system was replaced by a digital control system based on a then new device—the minicomputer. The Digital Equipment Corporation PDP–8 Minicomputer was the first of its kind and Code and his associates put it to very good use in one of the first real-time digital control systems of any sort. This APT worked well for the purpose intended, observing very bright standard stars to determine extinction, and many of the participants in the University of Wisconsin program went on to numerous accomplishments in space astronomy. This system was described in some detail by McNall, Miedaner, and Code (1968) and details on the developmental history are given in Genet (1986). This successful pioneering effort was the clear predecessor of those that were to come later.

In a parallel effort, Aden Baker Meinel, then Director of Kitt Peak National Observatory, and his associates formed a Space Division and, among other things, developed an APT and used it to make the first automatic observations of variable stars. They procured a 50–inch Boller and Chivens telescope and installed it on Kitt Peak. A Packard–Bell computer was installed in their downtown Tucson headquarters. Stephen Maran later took charge of the project and, after many years of effort, "Automated Photometry of the Variable Star 14 Aurigae" was published in the *Astrophysical Journal* (Hudson et. al. 1971). The complexity of the system and the poor reliability of the computer precluded regular operation of the system and the 50–inch telescope soon thereafter was converted to manual operation. While this APT was at best a limited success, Maran (1969) clearly foresaw the usefulness of future APT's for routinely observing variable stars, even though this capability had not yet arrived. An effort somewhat similar to that of KPNO was the automation of a 24–inch Boller and Chivens at Michigan State

University by Albert P. Linnell. A Raytheon minicomputer was used and the code was written in editor/assembler. As with the KPNO effort, after a single light curve was published, the system's automatic functions were discontinued. A very different and interesting effort was initiated by Stirling Colgate and his associates at New Mexico Institute of Mining and Technology. This system involved, as did the KPNO effort, the difficulties of both automation and a remote computer. It was developed for an automatic supernova search. The project is into its second decade of development. A later automatic supernova search program by the Lawrence Berkeley Laboratories has succeeded.

While not fully automatic, because it required manual startup and shutdown as well as an operator at all times, the 48–inch system at Cloudcroft, New Mexico did make routine observations of solar-type stars in clusters with considerable success for several years until the system was dismantled. This effort by Richard R. Radick and his associates used a telescope developed by the United States Air Force for the photometry of satellites. The operator placed the telescope on the first star to be observed in a cluster and, after that, the system would automatically find, center, and measure the other stars in the cluster. At the end of the fixed sequence, the system would alert the operator who would then move the telescope manually to the first star of the next cluster. This semiautomatic system was a clear success because it produced useful scientific output over a period of time. It was unfortunate that the telescope and antiquated IBM 1800 computer were so expensive to maintain that their continued operation was not economical.

Another semiautomatic system was David Skillman's. As was the case with the Cloudcroft system, the operator had to place the telescope on the first star manually but, unlike the Cloudcroft system, Skillman could and did routinely go to sleep and leave his telescope to continue observations all night, unattended. Although such unattended operation meant that observations must continue on the same variable star (and comparison star and sky positions), this was not a serious handicap. The system would be placed on a variable star in the east (usually an eclipsing binary going into eclipse) and, while Skillman slept, observations would continue until the variable set in the west, dawn came, or the weather turned bad. Two microcomputers, a KIM and an Apple, controlled the 12-1/2–inch telescope and its photometer, which is shown in Figure 7–1. The system has been quite successful. Observations on some 40 eclipsing binaries have been published in a paper in the *JAAVSO* and another paper, to be published, presents over 100 observational hours on V711 Tau. This system has been described briefly by Skillman (1981, 1982), and its development by Genet (1986).

Certainly the most productive APT to date has been the one developed by Louis J. Boyd. This fully automatic system has been observing a program of some 80 variable stars (and a similar number of comparison and check stars) for over five years, as this Second Edition went to press. The combination of full automation and clear Arizona skies has produced a mountain of high quality data and several dozen papers. While perfectly capable of devoted observing of a single short-period variable star, this system has been used mainly to make once-a-night measurements of *many* variable stars of medium to long period. These stars need only be loaded once on the master list and they will then be observed automatically for

**Figure 7–1.** `The Skillman semiautomatic telescope.`

as long as desired—typically years. Many of these stars were RS CVn binaries and one of us (Hall) was the primary recipient of this early APT data. The development of this system has been described by Genet (1986). The "Phoenix APT" developed by Louis Boyd is shown, together with its developer, in Figure 7–2.

A much simplified and improved second-generation Boyd–type APT was developed as a team effort directed by one of us (Genet). DFM Engineering designed the telescope mount and drive system (Melsheimer and Genet, 1984), Meade Instruments provided the Schmidt Cassegrain 10–inch optical assembly, Optec Inc. provided the SSP–3 solid state photometer, and the control system was based on a design by Louis Boyd. This system was described in great detail by Trueblood and Genet (1985), including complete wiring schematics and software listings. This system first became operational in Ohio in 1984, but was soon moved to Arizona where, after only slight reworking, it became the first operational APT of the Automatic Photoelectric Telescope Service on Mt. Hopkins.

## 7.3   How an APT Works

The operation of an APT perhaps can be understood best by considering its normal sequence of operations. By way of example, we will consider the operational sequence for the second generation APT system as described by Boyd et al. (1984).

**Figure 7–2.**  *Louis Boyd and the Phoenix APT*

During the day, procedure TILDARK checks the time and calculates the position of the sun. As soon as astronomical twilight occurs, the main control procedure is called. MAIN then calls procedure STARTSCOPE, which places the telescope in a known initial starting position by moving the telescope until it trips optical limit switches in the southeast corner of the telescope's travel range, its "home" position. STARTSCOPE also initializes the filter wheel and turns on the tracking rate generator so that the telescope starts moving toward the west.

Procedure MAIN then selects which "group" of stars to observe first. A group consists of variable, comparison, and check stars and a sky position (usually midway between the variable and comparison). There is a mathematical "observing window" above the telescope within which all observing must be done. The window can be set to avoid pointing at trees, going too close to the horizon, exceeding telescope's physical travel limits, etc. The group chosen for observation is the first one that will set on the western edge of the observing window (and which is at least 10° away from the moon).

This selection is made from among all the groups in the entire observing program; thus an observing program need be loaded only once and the system itself will choose what stars should be observed and in what sequence. Because the group that will set first on the western edge of the observing window is the first to be observed, groups will be observed from west to east and dawn will find the system working groups just as they rise over the eastern edge of the observing window. This approach maximizes the coverage of groups over the year. Other strategies are possible, such as observing groups closest to the meridian.

Procedure MOVE is then called to move the telescope to the group to be observed. A move is made in two segments with an intermediate stop, as shown in Figure 7–3. This allows a single source of ramped pulses to move the telescope with both right ascension and declination steppers operating simultaneously during the first segment. After a fleeting stop between the two segments, a single stepper is operated in the second segment to bring the telescope to its final destination.

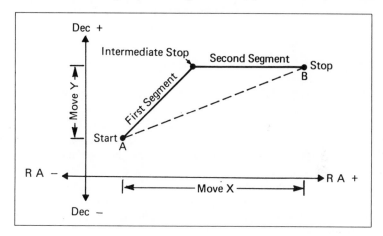

**Figure 7–3.** *How the command* MOVE *positions the telescope to a star group.*

Experience shows us that, even after traveling clear across the sky with an open-loop stepper system (there are no optical shaft encoders), the telescope is always within a few arcminutes of the desired position, which is the check star in the group. Before a search can be initiated for the check star, it is necessary to estimate how bright the check star will be, so an appropriate threshold can be set. This estimate is based on the sky brightness, the expected extinction (as a function of zenith angle), and the catalog brightness of the star. The sky brightness is determined by making five measurements in a "square with center" pattern and taking the lowest one.

Procedure HUNT is then called. It executes the "square spiral" search pattern shown in Figure 7–4. Starting in the center of the spiral, it makes a 0.2–second integration and compares the value obtained with the previously set threshold. If it is not exceeded, the telescope is moved to the next position. This continues until the star is found or the 12th spiral has been completed, whichever occurs first. If a star is found, it is centered using procedure LOCK, described below. If the star is not found (which experience shows us happens only if the sky has become cloudy), then the next group is selected, the telescope is moved to it, and another search is initiated. If it, also, is not found, then STOPSCOPE is called and the system shuts itself down automatically.

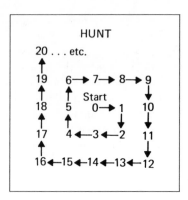

**Figure 7–4.**   The "square spiral" search pattern of HUNT procedure.

It might be thought that, with so many stars in the sky, the wrong star might be acquired on occasion. In five years of operation, the system in Phoenix has successfully acquired 1,000,000 stars. A crucial factor responsible for this success is the selection of the check star, which we sometimes refer to as the "navigation star." The check star is chosen to be in an optimal magnitude range for the system and to be clear of any nearby and/or comparably bright stars that might confuse the system. After the long move to a check star and the acquisition of it, the moves to the comparison star and the variable star are over short distances. Here the accuracy of the open-loop telescope positioning is very good, usuallyally better than an arcminute.

Once the star is acquired, it is centered in the diaphragm by procedure LOCK. Because single "X" and "Y" scans were found to be slow and inefficient, we use an iterative procedure that requires just four measurements on each iteration. From the initially assumed center position, the telescope is moved to four different positions which allow a slight overlap of the diaphragm edges over the assumed center position, as is shown in Figure 7–5. At each of these four positions a check is made for a star that exceeds the threshold. If a star is detected at all four positions, then the star must be centered and the job is finished. If the star shows up at both left positions but neither right position, then the telescope must be moved to the left. Other directions can be deduced from other combinations. Of the 16 combinations possible, two are logically illegal (opposite corners); if either of these is detected, the process is repeated. After each set of four measurements, the telescope is moved to a new assumed center not far from the original one, but in the correct direction. After just a few iterations, our experience shows, the star is always centered.

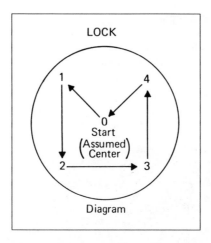

**Figure 7–5.** *Centering a star in the diaphragm with procedure* LOCK

With the first (check) star centered, measurements of 10 seconds duration each are made in the desired bands such as U, B, and V. The telescope is then moved to the comparison star position and that star is acquired, centered, and measured. The complete sequence of moves and measurements made to complete the 33 observations in a group, when doing UBV photometry, is shown in Table 7–1. This yields three bracketed differential measurements (each in U, B, and V) between comparison and variable. These differential measurements are examined in real time for internal consistency. If this examination is failed, then the group is remeasured. If it is passed, then the next group is selected and observed.

| | TABLE 7–1 | | | |
| | SEQUENCE OF | | | |
| | MOVES AND MEASUREMENTS | | | |
| MOVE | POSITION | U | B | V |
|---|---|---|---|---|
| 1 | Check Star | 1 | 2 | 3 |
| 2 | Sky | 4 | 5 | 6 |
| 3 | Comparison Star | 7 | 8 | 9 |
| 4 | Variable Star | 10 | 11 | 12 |
| 5 | Comparison Star | 13 | 14 | 15 |
| 6 | Variable Star | 16 | 17 | 18 |
| 7 | Comparison Star | 19 | 20 | 21 |
| 8 | Variable Star | 22 | 23 | 24 |
| 9 | Comparison Star | 25 | 26 | 27 |
| 10 | Sky | 28 | 29 | 30 |
| 11 | Check Star | 31 | 32 | 33 |

This process is repeated until the system either fails to find two groups (almost always because of clouds) or it determines from the calculated sun position that astronomical dawn has occurred. In either event, the system is returned to its home position and the roof closed.

## 7.4   The Design of APT's

The optics and mounts for the early APT's tended to use either somewhat conventional telescopes or telescopes designed for other purposes (such as extraterrestrial space). As design work began on larger APT's, it was realized that there were both special requirements and special opportunities in the case of totally dedicated (photometry) and totally automatic telescopes. An example of such a design is the 0.75–meter APT at the APT Service, shown in Figure 7–6.

**Figure 7–6.** *The 0.75–meter APT telescope as installed at the APT Service.*

## 7.5   The Automatic Photoelectric Telescope Service

The natural economies of APT operations are greatly magnified when a number of APT's are grouped together at a single site. This "economy of scale" results from savings in a number of areas. One economy of scale is manpower. While APT's do not require any observers, they do require custom development, fabrication, installation, operation (including the consolidation and reduction of data), and maintenance. All these activities require that considerable knowledge of APT's be acquired and maintained. Once up to speed on one APT, a single person can apply this knowledge to a number of similar systems if they are grouped together. We estimate that one or two persons can handle ten similar, co-located APT's, while it takes one person about half time to handle a single APT. Another significant economy for multiple APT's comes from the use of common facilities and equipment. These include: the weather sensors, roof control, power conditioners, computer warm area, road access, power, telephones, electronic test equipment,

**Figure 7-7.** *Mt. Hopkins in Southern Airzona.*

tools, etc. The requirements for all of these are not related to the number of APT's supported.

Recognizing the considerable advantages of grouping APT's together, the Automatic Photoelectric Telescope Service was founded in January 1985. The APT Service is located at the 7800-foot level on Mt. Hopkins in southern Arizona and is managed by the Fairborn Observatory (equipment, operations, and maintenance) and the Smithsonian Institution (buildings and site) on behalf of a number of participating institutions. Two basic types of participation in the APT Service are open to institutions. In the first, the participating institution owns and completely controls its own APT. The APT Service installs, operates, and maintains the APT for the institution that owns it. The cost for doing this is modest because of the many efficiencies of group APT operation. The first institution to utilize this form of service was Vanderbilt University with a 0.4-meter APT devoted to photometry of RS CVn binaries.

In the second, the participating institution rents time on one of the APT's owned by the APT Service itself, sharing this APT with other institutions also renting time on it. If the number of observations needed on it is not great and one of the existing telescope/filter/detector combinations owned by the APT Service is adequate, then this service can be helpful and low in cost. Observations gathered on a star in this manner typically cost less than the page charges for publishing the results! The first institutions to use this service were the University of Toronto, Franklin and Marshall College, Wesleyan University, and the University of Arizona. In addition to these, a number of other institutions have now instituted operations with the APT Service.

**Figure 7–8.**  *The first three APT's on Mt. Hopkins.*

The APT service can be found near the end of a twisty dirt road some 20 miles long (Figure 7–7). It is located in two buildings, one of which contains the control computers, shop, office, and sleeping space. The other building has a large roll-off roof and houses the telescopes. In this second building there is sufficient room for about ten APT's, including several of 0.75–meter size. The first three APT's on Mt. Hopkins are shown together in Figure 7–8. They are located in one corner of the roll-off-roof building and take up less than one-fourth of the total floor space.

**Figure 7–9.**  *Russell Genet (left) and Louis Boyd with the Fairborn–10 APT.*

**Figure 7–10.** *Douglas Hall with the Vanderbilt 16–inch APT on Mt. Hopkins.*

The Fairborn–10 APT was installed in September 1985. Figure 7–9 shows Genet (left) and Boyd standing behind this instrument with the multiple mirror telescope (MMT) in the background. The Vanderbilt 16–inch was the second APT at Mt. Hopkins. This telescope and mount were built by DFM Engineering. The control system and photometer were built by the Fairborn Observatory. Funded by the National Science Foundation, this telescope is at present devoted to photometry of spotted stars in RS CVn–type binary systems. Hall, the principle investigator, is shown in Figure 7–10 standing beside this APT. The Phoenix–10 APT was relocated to its new home on Mt. Hopkins after a freak thunderstorm blew the roof off its enclosure in Phoenix. Like its namesake, it has risen from its ashes to fly again.

# Chapter 8

# PHOTOELECTRIC OBSERVATORIES

## 8.1  Introduction

In Chapters 3 through 6 we have discussed the various elements of photoelectric systems. This chapter gives actual examples of a variety of ways in which the pieces can be assembled into operating systems. The remainder of the book suggests how a complete system might actually be used.

No two photoelectric observatories are the same, and each observatory has worked out a unique solution to its particular circumstances of location, finances, personal desire, etc. Faced with such variety, it is only natural to develop some classification scheme—even if somewhat arbitrary.

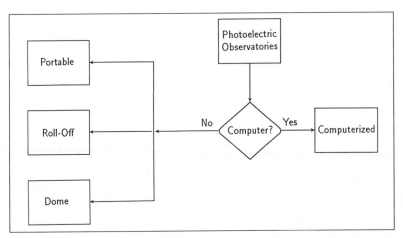

**Figure 8–1.**  *Classification of photoelectric observatories.*

To present the observatories in this chapter, we have classified them as shown in Fig. 8–1. The limited space has precluded extensive discussions, so we concentrated on pictures kindly provided by the individual observatory proprietors, hoping that each picture will be worth a thousand words. The name and address of each observatory has been included so that interested readers can write for more details. We regret that space and time did not allow inclusion of the

many other fine smaller observatories around the world that are engaged in photoelectric observations of variable stars. Detailed descriptions of numerous other
small-telescope observatories equipped for photoelectric photometry have, however, been published elsewhere. Table 8–1 is a listing of such descriptions which
have appeared in *Advances in Photoelectric Photometry (Volume I and II)*, in
*Micro Computers in Astronomy (Volume I and II)*, and in the *IAPPP Communications*.

Among the smaller observatories, those without a direct connection between
the photometer and a computer are in the vast majority. This is as it should
be because, in most cases, the expense and bother involved in making a direct
photometer-computer connection is simply not worthwhile and does not contribute
to the quality of the observations.

The portable photoelectric observatory is the least expensive, most flexible,
and fastest growing type of observatory. The readily available and highly compact
Schmidt–Cassegrain telescopes are almost ideal for small-observatory photometry,
and the new compact heads and electronics are very compatible with portable
telescopes. Getting started with productive photoelectric photometry (making
precision observations of variable stars) can be an easy, simple, quick, three-step
process. Buy a Schmidt-Cassegrain telescope from Meade (8–inch or 10–inch)
or Celestron (C–8, C–11, or C–14). Obtain one of the comercially available self-
contained photometers (from Optec or EMI Gencon). Add a small pier or concrete
pad in your backyard with pre-aligned slots for your telescope. And now you are
ready to start making your contribution to science.

While the portable photoelectric observatory is without question the most
cost-effective, a permanent home for the telescope, photometer, and associated
paraphenalia is an attractive luxury where circumstances permit. A slightly over-
sized "garden shed" with a roll-off or fold-off roof makes a good, low-cost observatory. In a residential setting, such an unobtrusive structure avoids problems with
zoning, building inspectors, irate neighbors, and curious passers-by. No one need
know that the "shed" does not really contain a riding mower, but instead contains
a complete astronomical observatory used to make precise scientific measurements
of variable stars!

While a roll-off or fold-off roof provides the most economical and easily operated permanent observatory structure, a dome is the traditional roof for an
observatory. Besides providing excellent protection against wind and light, the
dome has a symbolic significance and attractiveness that is almost irresistible.
Who is there among us who would not like to have their own observatory on a
secluded mountain with a bright, shiny dome?

The computerization of photometry at smaller observatories is still in its in-
fancy and, while its future looks bright, it is not entirely clear at the time this was
written whether, in the long run, this technology will become widespread among
smaller observatories. The computerized systems at the Fairborn Observatory
have already been discussed in some detail in earlier chapters, so the information
on them will not be repeated here.

| TABLE 8–1, Part 1 | | | | | | |
|---|---|---|---|---|---|---|
| Small-telescope Observatories Equipped for Photoelectric Photometry | | | | | | |
| IAPPP = I.A.P.P.P. Communications | | | | | | |
| APP = Advances in Photoelectric Photometry, Volumes I and II | | | | | | |
| MCA = MicroComputers in Astronomy, Volumes I and II | | | | | | |
| Observatory Name | Observatory Location | Observatory Proprietor | Aperture (inches) | IAPPP (No.) | APP (Ch.) | MCA (Ch.) |
| Auckland | Auckland New Zealand | W.S.G. Walker | 20 | 21 | | |
| Stephen F. Austin State University | Nacogdoches Texas | J.B. Rafert N.L. Markworth | 18,41 | | | I:6,14 II:9,10 |
| APT Service | Mt. Hopkins Arizona | L.J. Boyd R.M. Genet | 10,10,16 | 22 | | |
| Beverley Begg | Dunedin New Zealand | W.H. Allen | $8,12\frac{1}{2}$ | 2 | | |
| Big Cottonwood | Salt Lake City Utah | J.L. Foote | 20 | 25 | | |
| Borlik | Indianapolis Indiana | T. Borlik | 14 | | | II:15 |
| Boyd | Miami Florida | R.W. Boyd | 8 | 6 | | |
| Braeside | Flagstaff Arizona | R.E. Fried | 16 | 16,25 | | I:8,12,24 II:5,6 |
| College de Levis | Levis, Que. Canada | A. Tardif | 8,14 | 10 | | |
| Dance Hill | Kitchener, Ont. Canada | C. Cunningham M. Kaitting | $14\frac{1}{4}$ | 8 | | I:26 |
| DeWitt | Nashville Tennessee | J.H. DeWitt | 12 | 19 | | |
| Douglass | South Miami Florida | W.T. Douglass | 14 | 14 | | |
| Droke | Fayetteville Arkansas | C.P. Lacy | 16 | 25 | | |
| East Tennessee State Univ. | Johnson City Tennessee | H.W. Powell J.R. Miller | 8 | 23 | | II:16 |
| Einstein Astrophysical | Sidney Ohio | J.C. Soder | 10 | 21 | | |
| Fairborn East | Fairborn Ohio | R.M. Genet | 8,10,16 | 3,9 | II:6 | I:7 II:3 |
| Fairborn West | Phoenix Arizona | L.J. Boyd | 10 | 15 | II:6 | II:3 |
| Florida Institute of Technology | Melbourne Florida | T.D. Oswalt J.B. Rafert | 14 | 20 | | |
| Fulton | Garden City Kansas | G.A. Bower | 8 | 8,14 | | |
| Gettysburg | Gettysburg Pennsylvania | L.A. Marschall | 16 | | | II:18 |

TABLE 8–1, Part 2
Small-telescope Observatories Equipped for Photoelectric Photometry

IAPPP = I.A.P.P.P. Communications
APP = Advances in Photoelectric Photometry, Volumes I and II
MCA = MicroComputers in Astronomy, Volumes I and II

| Observatory Name | Observatory Location | Observatory Proprietor | Aperture (inches) | IAPPP (No.) | APP (Ch.) | MCA (Ch.) |
|---|---|---|---|---|---|---|
| Gila | Claypool Arizona | K.W. Zeigler | 11 | 14 | | |
| Grim | Stansbury Park Utah | B.S. Grim | 16 | 14 | | |
| Grundy | Lancaster Pennsylvania | M.A. Seeds D.W. Dawson | 16 | 11 | II:4 | |
| Grup d'Estudis Astronomics | Barcelona Spain | E. Melendo | 8,16 | 20 | | |
| Hickox | Chagrin Falls Ohio | L.P. Lovell | 10 | 6 | | |
| Hillside | Cedar Falls Iowa | D.J. Hoff P.S. Leiker | 16 | 26 | | |
| Hopkins-Phoenix | Phoenix Arizona | J.L. Hopkins | 8 | 9 | | I:17 II:19 |
| Indian River | Ottawa, Ont. Canada | B. Burke | 16 | 2 | | |
| Johnson | Mitchellville Maryland | J. Johnson | 11 | | I:17 | |
| Krisciunas | San Jose California | K. Krisciunas | 6 | 6 | | |
| Lake Afton Public | Wichita Kansas | D.R. Alexander | 16 | 14 | | |
| Lewis and Clark | Bozeman Montana | W.E. Newsome | 11 | 24 | | |
| Limber | San Antonio Texas | D. McDavid | 14 | | | II:14 |
| Lines | Mayer Arizona | R.D. Lines H.C. Lines | 20 | | I:19 | |
| Manuel de Barros | Vila Nova de Gaia Portugal | L.M. Ribeiro J.F. Carneiro | 30 | 21 | | |
| Martin | Sevilla Spain | A.M. Martin | 8 | 14 | | |
| Mercyhurst College | Erie Pennsylvania | R.C. Reisenweber | 8 | 21 | | |
| Minot State College | Minot North Dakota | D. Martin | 16 | 23 | | |
| Monash | Clayton, Victoria Australia | D.W. Coates K. Thompson J.L. Innes T.T. Moon | 16,18 | | II:7 | |
| Mouldsworth | Cheshire England | R. Miles | 11 | 18 | II:14 | |
| Mount Tamborine | Brisbane Australia | A.A. Page | 12 | 10 | | |

| TABLE 8–1, Part 3 Small-telescope Observatories Equipped for Photoelectric Photometry | | | | | | |
|---|---|---|---|---|---|---|
| IAPPP = I.A.P.P.P. Communications APP = Advances in Photoelectric Photometry, Volumes I and II MCA = MicroComputers in Astronomy, Volumes I and II | | | | | | |
| Observatory Name | Observatory Location | Observatory Proprietor | Aperture (inches) | IAPPP (No.) | APP (Ch.) | MCA (Ch.) |
| Nessa | Nessa East Germany | D. Böhme | $6\frac{1}{2}$ | 13 | | |
| Nielsen | Wilmington Delaware | P. Nielsen R. Kellerman | 4 | 12 | | |
| Nigel | Nigel South Africa | L. Pazzi | $12\frac{1}{4}$ | 21,27 | I:15 | |
| N. Valley Stream | N. Valley Stream New York | F.J. Melillo | 8 | 26 | | |
| Ormada | Cheshire England | A.J. Hollis | 5,12 | 16 | | II:17 |
| Pulpit Rock | Allentown Pennsylvania | R. Wasatonic | 20 | 8 | | |
| Rolling Ridge | Erie Pennsylvania | R.C. Reisenweber | 8 | | I:14 | |
| Scuppernong | Dousman Wisconsin | T.R. Renner | 10 | 8 | | |
| Shenandoah | Hopewell Junction New York | T.G. McFaul | 8,14 | 6 | | |
| Slote | Fire Island New York | S. Slote | 8 | 14 | | |
| Stanton | Joshua Tree California | R.H. Stanton | 16 | | I:9 | |
| Stelzer | River Forest Illinois | H.J. Stelzer | 14 | 8,14 | | |
| Stokes | Hudson Ohio | A.J. Stokes | 6 | 26 | | |
| Sunset Hills | Hacienda Heights California | N.F. Wasson | 8 | 21 | | |
| Tardis | Waterloo, Ont. Canada | M. Kaitting | $14\frac{1}{4}$ | | | I:15 |
| Tiara | Colorado Springs Colorado | T.E. Schmidt | 16 | 14 | | |
| Tjorn Island Astronomical | Skarhamn Sweden | S.I. Ingvarsson | 14 | | I:16 | |
| Univ. of Utah | Salt Lake City Utah | K.D. Green | 16 | 14 | | |
| Van Vleck | Middletown Connecticut | W. Herbst | 24 | 11 | II:9 | |
| Wheaton College | Norton Massachusetts | T. Barker | 14 | 25 | | |
| Winer Mobile | Potomac Maryland | M. Trueblood | 14 | 22 | | |

## 8.2 Portable Observatories

### 8.2.1 The Fulton Observatory

This observatory is living proof that photoelectric photometry does not have to be expensive or complicated. The telescope is a commercially made 8–inch $f/6$ Newtonian and the photometer is a solid-state Optec unit. The observatory belongs to high school student Gary Bower, 1203 Labrador Blvd., Garden City, KA 67846. Bower's complete observatory is shown in Photograph 8–1. He observes the brighter eclipsing binaries with the same precision as the largest observatories but, unlike them, he does not have to stop down his telescope to avoid overloading his photometer! As the second edition of this book went to press, Bower had graduated *summa cum laude* with a major in astrophysics from the University of Oklahoma.

### 8.2.2 The Stelzer Observatory

Without question, Harold Stelzer, 1223 Ashland Avenue, River Forest, IL 60305, has the most nicely equipped portable photoelectric observatory in the world. His Celestron C–14 is mounted on a sturdy mount which he rolls out of his garage to a small concrete pad in his backyard. A few turns of three knobs and presto— he has a very solid, perfectly aligned 14–inch telescope as shown in Photograph 8–2. A roll-out cart contains all the electronics, charts, etc. For windy nights he even has portable wind screens that fasten into pre-established posts. What is particularly amazing is that Stelzer, from a dead start, can be observing while most astronomers are still rolling their roofs off or adjusting their domes. Equally amazing, from his city location near Chicago, Stelzer observes many of the fainter eclipsing binaries with an accuracy consistently better than 0.01 magnitude.

### 8.2.3 The Tretta Observatory

The observatory of Fred Tretta, 6316 W. Sarrey Street, Glendale, AZ 85304, is not portable in the usual sense. Mr. Tretta's 10–inch Newtonian telescope, shown in Photograph 8–3, remains outside most all the time and, when not in use, is covered with a tarpaulin. The fine Arizona weather makes this a practical approach. The Optec photometer is used primarily to make VRI (visual, red, infrared) observations of late-type variable stars. Mr. Tretta is one of the few amateur astronomers engaged in infrared photometry, and the high quantum efficiency of the photodiode in his Optec photometer in the near IR is well-suited to this task.

### 8.2.4 The Johnson Observatory

The final example of a portable observatory is that of Jaurvon Johnson, 3305 Meadowridge Place, Mitchellville, MD 20716. Mr. Johnson's observatory is stored in his garage when not in use and can be quickly transported with a modified dolly to the observing site in the backyard (Photograph 8–4). Not

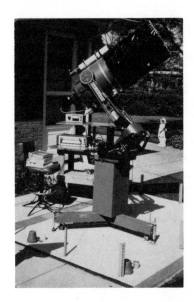

**Photograph 8–1. (Left)** *Telescope and photometer at the Fulton Observatory. Not shown is a sandbag used to steady bottom of the mount.*

**Photograph 8–2. (Right)** *Stelzer portable observatory. Roll-out cart on the left. Angle irons hold the wind screens (not in place).*

**Photograph 8–3. (Left)** *Tretta Observatory. Optec photometer head can be seen on scope at the left. The electronics are on the right.*

**Photograph 8–4. (Right)** *Mr. Johnson at the controls of his observatory.*

shown is a Radio Shack TRS–80 computer that is used to reduce and analyze the data. Mr. Johnson even has a program that plots and labels his light curves. An Epson MX80 printer is used for this.

## 8.3   Roll-Off Roof Observatories

### 8.3.1   The Hopkins-Phoenix Observatory

The observatory of Jeffrey Hopkins is located at 7812 W. Clayton Drive, Phoenix, AZ 85033. The 8–foot square observatory houses a Celestron C–8 on a very solid pier as shown in Photograph 8–5. The lightweight roof is easily rolled off with one hand. Observations, sometimes assisted by daughter Stephanie, are primarily of eclipsing binaries of the RS CVn and Zeta Aurigae types. Besides his observational program, Hopkins is active in the design and construction of photoelectric equipment and in the various activities of IAPPP. Hopkins was co-director (photometry) of the International Epsilon Aurigae Campaign.

### 8.3.2   The Brettman Observatory

The observatory of Orville Brettman, 2730 N. Elston Avenue, Chicago, IL 60647, houses a 12–inch Cassegrain telescope equipped with an EMI Gencom STARLIGHT–1 pulse-counting photometer. The telescope and photometer are shown in Photograph 8–6. A Radio Shack TRS–80 is used for data reduction and analysis. In the very first series of photoelectric observations from his new observatory, Brettman discovered a new variable star: HR 5 (Brettman et al. 1983).

## 8.4   Domed Observatories

### 8.4.1   The Louth Observatory

The observatory of Howard Louth is located at 2199 Hathaway Road, Sedro Wool-ley, WA 98284. The beautifully machined 10–inch telescope and photometer can be seen in Photograph 8–7. Note the generous-sized finder. The electronics, shown at the left of the picture, include a pulse counter, digital clock, and printer. Mr. Louth is a veteran observer with many published papers to his credit.

### 8.4.2   The Lines Observatory

The observatory of Helen and Richard Lines, 6030 North 17th Place, Phoenix, AZ 85016, is located at 4500–foot elevation between Phoenix and Flagstaff. The photometer at the Lines Observatory, shown in Photograph 8–8, is dwarfed by the 20–inch Cassegrain telescope. One of the brightest known eclipsing binaries, 5 Ceti, was discovered after patient and highly precise photoelectric observation at the Lines Observatory (Lines and Hall 1981).

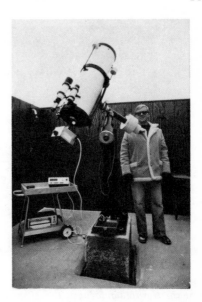

**Photograph 8–5.** **(Left)** *The Hopkins-Phoenix Observatory. The Celestron C-8 is used with photoelectric equipment built by Mr. Hopkins to observe RS CVn binary stars.*

**Photograph 8–6.** **(Right)** *The observatory of Orville Brettman. The 12–inch Cassegrain telescope and SN#1 EMI Gencom STARLIGHT–1 photometer.*

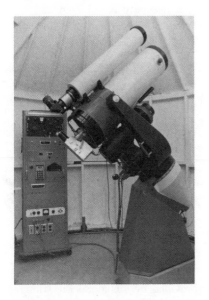

**Photograph 8–7. (Left)** *The 10–inch telescope at the Louth Observatory. The pulse counting, and auxiliary electronics are all in the rack at the left.*

**Photograph 8–8. (Right)** *Richard Lines at the business end of his 20–inch Cassegrain telescope.*

**Photograph 8–9.**  *The Renner telescope in its stowed position. The photometer head is at the Newtonian focus at the left and the electronics are by the pier.*

### 8.4.3  The Scuppernong Observatory

The observatory of Thomas R. Renner, 4512 Deerpark Drive, Dousman, WI 53118, is shown in Photograph 8–9. The 10–inch telescope is equipped with a home-built UBV photometer. The dome is more than a necessity in the cold and windy Wisconsin winters. Mr. Renner is a very experienced observer of eclipsing binaries with a number of published papers to his credit.

**Photograph 8–10a and b.**  *The Tiara Observatory at Eleven-Mile Canyon.*

### 8.4.4  Tiara Observatory

The 16–inch Ealing Cassegrain telescope (provided by a General Electric grant) and Optec photometer of the Colorado College are shown in Photographs 8–10a and b. Located at 8800–foot elevation as part of Terry Schmidt's observatory, it is one of the highest small observatories equipped for photoelectric photometry. Dr. Charles Bordner, Department of Physics, The Colorado College, Colorado Springs, CO 80903, is the Director of the observatory.

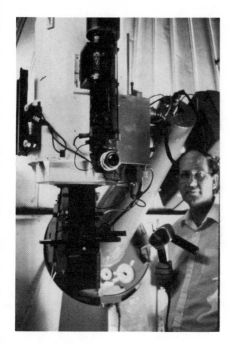

**Photograph 8–11.** (**Left**) *Coauthor Russ Genet on the 16–inch telescope at Kitt Peak National Observatory.*

**Photograph 8–12.** (**Right**) *The U.S. Air Force Academy 24–inch telescope and photometer.*

### 8.4.5  KPNO No. 3 16–Inch telescope

At Kitt Peak National Observatory the No. 3 16–inch telescope was one of the first telescopes on the mountain and has been a photometric workhorse for two decades. A mountain of photoelectric data has been provided by this one small telescope over the years. The Boller and Chivens telescope is shown in Photograph 8–11 (photograph by George C. Roberts). One of us (Genet) is at the controls during an observing run with Kenneth Kissell and George C. Roberts. It is with some sadness that we note this highly productive telescope has now been shut down.

### 8.4.6  U.S. Air Force Academy Observatory

The then Director of the USAFA Observatory, Captain Wayne Hanson, Department of Physics, U.S. Air Force Academy, CO 80840, kindly supplied a photograph of the 24–inch Boller and Chivens telescope and UBV photometer. The photometer was built at the academy and is based on the Lowell Observatory design. These are shown in Photograph 8–12. The current Director, Major Raymond Bloomer, has made many improvements in the photometric system.

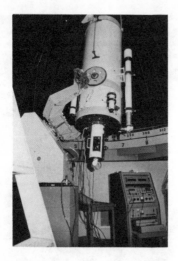

**Photograph 8–13. (Left)** *The 16–inch Cassegrain telescope at the Braeside Observatory.*

**Photograph 8–14. (Right)** *The 24-inch telescope at the Capilla Peak Observatory.*

## 8.5    Computerized Observatories

### 8.5.1    Braeside Observatory

West of Flagstaff, Arizona is Mars Hill, home of the world famous Lowell Observatory. The hill north of Mars Hill is known among small observatory photometrists as "Fried Hill." On its crest is the beautiful, domed Braeside Observatory. Under the dome is a 16–inch Cassegrain telescope and in the heated control room is an Apple microcomputer. The computerized system, designed by Edward Mannery, logs the photoelectric observations and time and also controls stepper motors in the diaphragm and filter wheels, to give automatic movement to the sky background and the changing of filters. The telescope and photometer are shown in Photograph 8–13. The Director of the Braeside Observatory is Robert E. Fried, Braeside Observatory, P.O. Box 906, Flagstaff, AZ 86002. Since the first edition of this book, this pioneering system has achieved automatic operation.

### 8.5.2    Capilla Peak Observatory

Another computerized photometer controlled by an Apple microcomputer is the Capilla Peak Observatory of the University of New Mexico. Photograph 8–14 shows the photometer on the 24–inch telescope. The control program is unusually flexible and allows many observer decisions to be made prior to and even during the observing run. The office address is Dr. Michael Zeilik, Department of Physics and Astronomy, University of New Mexico, Albuquerque, NM 87131.

# Chapter 9

# THE EARTH'S ATMOSPHERE

If you were observing from an orbiting satellite or from an observatory on the moon, you could ignore this chapter. Unfortunately, though, one of the prices you pay for the convenience of a backyard telescope is the atmosphere which shrouds it and tries to stop the starlight during the last millisecond of its long journey from the star to the photomultiplier. The earth's atmosphere affects photoelectric photometry in several completely different ways. You are familiar with these effects if you are familiar with extinction, seeing, diffraction, dispersion, scintillation, and sky brightness.

Because brightness has no meaning unless the wavelength is specified, photometry must involve brightness measured at a specified wavelength or at several specified wavelengths. Although standard multi-bandpass photometric systems are discussed in Chapter 10, this chapter will, for the sake of illustration, refer frequently to the ultraviolet (u), blue (b), and visual (v) bandpasses of the well-known standard UBV photometric system.

## 9.1  Extinction

In describing the physics of the interaction between electromagnetic radiation (light) and matter, there is a distinction between absorption and scattering. In scattering, a photon "bounces off" an atom or particle, thereby suffering a change in energy and direction but not losing its identity. In absorption, a photon and all of its energy actually is absorbed by an atom or molecule. That atom or molecule may, after the passage of some time, reemit another photon (or even more than one photon) or it may shed the excess energy by some mechanism other than photon emission. In any case, the absorption ended the existence of the original photon. The stellar photoelectric photometrist knows only that starlight is dimmed or partially extinguished by passing through the earth's atmosphere, so we use the term extinction to describe the combination of scattering and absorption in whatever proportion they occur. The physicist uses the term opacity in a similar way. In yellow light, extinction can range from something like 15% at the zenith on a very clear night at a high-altitude observatory site, to virtually 100% when a night is cloudy.

If the transparency (or, more accurately, the semitransparency) of the atmosphere is uniform across the sky and holds constant with time, then it is a nuisance and compromises somewhat the accuracy of the photoelectric photometry but can

151

be dealt with. On the other hand, if the transparency is strongly variable across the sky and/or rapidly variable in time, then the sky is not photometric. One simply cannot do photometry on that night, not even differential photometry, unless one uses sophisticated two-channel photometric techniques such as are discussed in Chapter 6. Finely structured and rapidly moving cirrus is the clearest example of badly variable sky transparency which makes photometry impossible. A uniform layer of haze which might increase (or decrease) gradually throughout the night is an example of variable sky transparency which can, with care, be handled by the single-channel differential photometric techniques described in Chapter 9. Best of all, however, is sky transparency which is uniform across the entire sky and remains so throughout the entire night.

For most purposes in photometry, it is sufficient to idealize the earth's atmosphere as a plane parallel layer which is uniform in opacity both vertically and horizontally. Some theoreticians refer to this as the slab model of an atmosphere. It is not difficult to see how this is an idealization. The earth's atmosphere, like its surface, is not a plane but is spherically curved. The atmosphere has no well-defined edge, and its opacity is not constant with height but, along with its density, decreases from its maximum at ground level to smaller values at greater heights. Finally, to make matters worse, all of the above are strongly dependent on wavelength.

The slab model is illustrated schematically in Fig. 9–1. The thickness of the slab, measured vertically from the observatory on the ground to the top edge of the slab, is defined as one airmass. One airmass absorbs the same percentage of starlight as the real (curved, nonuniform, etc.) atmosphere does for a star at the observer's zenith. The so-called zenith distance of a star, actually an angle, is the angle between the observatory's zenith and a line connecting the star with the observatory. This is indicated in Fig. 9–1 with the symbol $z$. It is easy to see that the starlight's path through the slab is greater for non-zero zenith distances. Specifically, whereas the path length $X$ is one airmass at the zenith ($X = 1$ at $z = 0°$), it is greater by the factor $\sec z$ at other values of $z$. This gives us the equation

$$X = \sec z, \qquad (9.1.1)$$

which is called the secant approximation for airmass.

The value of $\sec z$ can be computed for any star from its equatorial coordinates with the expression

$$\sec z = (\sin \phi \sin \delta + \cos \phi \cos \delta \cos h)^{-1}, \qquad (9.1.2)$$

where $\phi$ is the observer's latitude, $\delta$ is the star's declination, and $h$ is the star's hour angle, which is – east of the meridian and + west of the meridian. Although it is not particularly difficult to compute $\sec z$ with modern electronic computers or even with pocket calculators, some observers find it useful to have a nomogram which lets one quickly estimate airmass graphically for a star of given $\delta$ and $h$. An example was given by Hardie (1962).

At large values of $z$ the secant approximation inherent in the slab model becomes increasingly worse. Long ago, Bemporad (1904) considered a model atmosphere which was still quite idealized but was more realistic than the slab

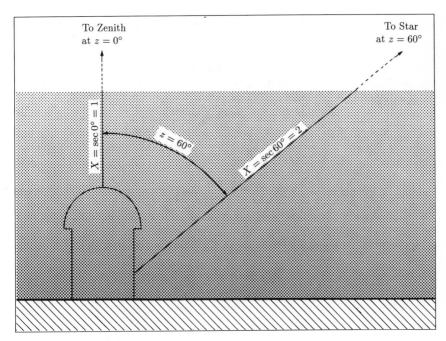

**Figure 9–1.** *Slab model schematic.*

model because it allowed for the atmosphere's curvature, the decrease in density with the height above the ground, and also the effects of atmospheric refraction. He derived values giving the difference between $X$ and sec $z$ for various values of $z$. Hardie (1962) used Bemporad's values to derive the useful interpolation formula

$$X = \sec z - A(\sec z - 1) - B(\sec z - 1)^2 - C(\sec z - 1)^3, \qquad (9.1.3)$$

where $z$ is apparent zenith distance, $A = 0.0018167$, $B = 0.002875$, and $C = 0.0008083$. Young (1974) considered various factors which can make even Bemporad's improved values inadequate at very large zenith distances. These factors include the fact that the real terrestrial atmosphere varies in opacity at various wavelengths and at various altitudes in a way considerably more complicated that Bemporad considered. Also, using local sidereal time to deduce hour angle and hence $z$, as many observers do instead of reading hour angle directly from their telescope, gives you true zenith distance rather than the apparent zenith distance which Bemporad's values presuppose. Snell and Heiser (1968) further note that any one expression for $X$, such as that in equation (9.1.3), will be applicable only to an observatory at a given height above sea level and that significantly different values of $A$, $B$, and $C$ would be needed for observatories at higher or lower altitudes above sea level. Young (1974) favors restricting most photometry to rather small values of $z$ in order to minimize all these problems. This advice would apply to all-sky photometry, to photometric determination of extinction coefficients, and to differential photometry. He shows that the simple formula

$$X = \sec z[1 - 0.0012(\sec^2 z - 1)], \qquad (9.1.4)$$

where $z$ is true zenith distance, works quite well up to $\sec z = 4$.

Because the physical process of absorption acts to absorb a certain *fraction* or *percentage* of the incident light, and because magnitudes are proportional to the *logarithm* of light intensity, it can be shown mathematically that the difference between the magnitude of the incident star light $m_o$ (before it encounters the earth's atmosphere) and the magnitude of the starlight reaching the telescope $m$ (after passing through the earth's atmosphere) is directly proportional to the air-mass $X$. The constant of proportionality is defined as the atmospheric extinction coefficient $k_\lambda$, where the subscript is inserted to emphasize that this constant is a function of wavelength. The expression is simply

$$m_o = m - k_\lambda X. \tag{9.1.5}$$

In a formal mathematical derivation it can be shown how $k$ is related to the density and the opacity of the earth's atmosphere. The term extraterrestrial magnitude is sometimes used to identify $m_o$ and the term raw magnitude to identify $m$. Notice that $X$ is always positive and $k$ is always positive and the minus sign is appropriate since the incident light must be brighter (a numerically smaller magnitude) than the transmitted light.

Atmospheric extinction is quite different depending on the wavelength of the light. The atmosphere is almost completely opaque at wavelengths shorter than about 3000Å  (the ultraviolet and shorter), at wavelengths longer than about 10 meters (the long-wave radio), and at many portions of the infrared spectrum between about 10,000Å  and 1 cm. Therefore, earth-based photometry of celestial objects must be carried out by observing through one of the atmosphere's windows, either the visible window, the radio window, or one of the several small infrared windows. The discussion which follows is limited to the visible window.

Even when the sky is "clear," the molecules and any particles (dust particles and/or water droplets) in the atmosphere cause extinction of starlight. The molecules absorb light at the wavelength invervals where molecular absorption bands occur. The molecules also scatter light at all wavelengths by a mechanism called Rayleigh scattering, which is roughly inversely proportional to the fourth power of the wavelength, i.e., stronger at shorter wavelengths.

The amount of extinction from Rayleigh scattering depends only on the wavelength and number of air molecules between the telescope and the star. You could actually calculate its strength if you knew the atmospheric pressure and elevation of the observatory. There is another component of extinction, however, which is virtually impossible to predict. Water droplet particles (present when there is haze in the air) and dust particles also scatter light at all wavelengths. Scattering by such small particles is called aerosol scattering. Particles with diameters comparable to the wavelength of light produce extinction which is roughly inversely proportional to the wavelength, while particles much larger than the wavelength of light produce virtually wavelength-independent extinction. With the relatively large water droplets the extinction is pretty well equal at all wavelengths. With the generally smaller dust particles it is stronger at shorter wavelengths. The overall result is that extinction decreases as one goes from ultraviolet wavelengths to infrared wavelengths. This is illustrated in Fig. 9-2, where $k$ at each wavelength has been derived from the values given by Allen (1973, §56) for the percent trans-

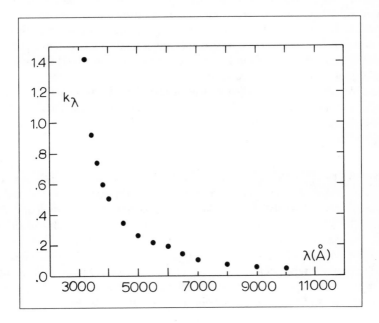

**Figure 9-2.**   *Extinction decreases as one moves from ultraviolet to infrared wave-lengths.*

mission through a normal clear atmosphere. It must not be forgotten, however, that the exact value of $k$ at each of the wavelengths of interest varies with observatory altitude, geographical location, and night-to-night sky conditions. For a particularly incisive discussion of extinction, see Hayes and Latham (1975).

Although the atmospheric extinction on a given night at a given observatory can be described in principle by a plot of $k$ versus $\lambda$, an unfortunate complication is introduced to broadband photometry because the bandpass defined by each filter is quite broad in extent and definitely not monochromatic. Thus, the value of $k$ appropriate to use for a given bandpass will be some sort of average of the $k$ versus $\lambda$ curve, taken over the wavelength interval of that bandpass. What makes the situation even more complicated is the fact that different stars have such different colors. To see the situation, let us imagine an idealized photomultiplier which is equally sensitive to all wavelengths and an idealized blue filter which transmits all light between 4000Å and 5000Å but is opaque elsewhere. Further, let us imagine three idealized stars: one which is "blue" and emits light only between 4000 and 5000Å, one which is much hotter and emits light only shortward of 4500Å, and one which is much cooler and emits light only longward of 4500Å. The appropriate value of $k$ to use for the blue star would be the average of the $k$ versus $\lambda$ curve over the 4000Å to 5000Å range. This will be pretty close to the value of $k$ at 4500Å but not exactly, because of the curvature of the $k$ versus $\lambda$ curve. The appropriate value of $k$ to use for the hotter star, though, would be an average of the $k$ versus $\lambda$ curve over the 5000 to 4500Å range, pretty close to the value at 4250Å. Similarly, the appropriate value of $k$ for the cooler star would be pretty close to the value at 4750Å. Thus, even on the same night and with the same blue filter, stars of different color will suffer different amounts of atmospheric extinction, the hotter stars relatively more and the cooler stars relatively less. *If* the $k$ versus $\lambda$ curve were flat *or* if the filters were nearly monochromatic *or* if stars emitted the same amount of light at all wavelengths, then this complication would not arise. But in reality, unfortunately, it does arise.

This complication is handled in the following way. As an example, let us consider a certain bandpass, $\lambda$. We consider the effective extinction coefficient at that bandpass to be composed of two parts and write

$$k_\lambda = k_\lambda' + k_\lambda''(B - V), \qquad (9.1.6)$$

where $k$ is the extinction coefficient, $k'$ is called the principal extinction coefficient, and $k''$ is called the color-dependent extinction coefficient. B–V is standard notation for the difference between the blue magnitude and the visual magnitude in the UBV system; as such it will be recognized as the color index of the star, a measure of the star's color. The UBV system defines B–V to be zero for stars of spectral type A0 which are not affected by interstellar absorption. Thus, stars of earlier spectral type, which are hotter and emit relatively more light in B than in V, will have a negative color index whereas stars of later spectral type, which are cooler and emit relatively more light in V than in B, will have a positive color index.

Combining equation (9.1.5) with equation (9.1.6) we get

$$m_o = m - k_\lambda' X - k_\lambda'' X(B - V). \qquad (9.1.7)$$

Obviously, $k'$ is that part of $k$ which is characteristic of the sky itself on the night in question, $k''$ depends on the width of the blue bandpass and the general shape (slope and curvature) of the $k$ versus $\lambda$ curve, and B–V depends on the color of the star being observed.

Hardie (1962) explains in detail how $k''$ can be determined for each bandpass in a given photometric system. For our purposes in differential photometry, however, we fortunately can make some intelligent approximations and effectively estimate values which are entirely adequate. The value of $k_v''$, when determined explicitly for most wide visual bandpasses, conveniently proves to be immeasurably small, i.e., within 0.01 of zero. Hence, it can be approximated as $k_v'' = 0^{\mathrm{m}}00 \pm 0^{\mathrm{m}}01$. The value of $k_b''$, when determined explicitly for most wide blue bandpasses, proves to be between about $- 0^{\mathrm{m}}02$ and $- 0^{\mathrm{m}}04$, depending primarily on the width of the filter. Hence, it can be approximated as $k_b'' = - 0^{\mathrm{m}}03 \pm 0^{\mathrm{m}}01$. Unfortunately, the value of $k_u''$, if the bandpass is broad and straddles the Balmer jump at 3647 Å, proves to be relatively large and a very strong function of the exact energy distribution of the star's light as a function of wavelength. It ranges between $0^{\mathrm{m}}00$ and $- 0^{\mathrm{m}}04$, depending on the star's spectral type and luminosity class and the amount of interstellar extinction which it has suffered. Because it is virtually impossible in practice to know enough about a star beforehand to determine what its value of $k_u''$ will be, the U magnitudes in the UBV system were set up by *assuming* $k_u'' = k_b''$, an expendient but really very poor assumption. This was, is, and always will be a basic flaw in the UBV system. Anyone else doing photometry which is to be transformed to the UBV system should, in order to match the UBV system as closely as possible, make that same poor assumption, i.e., that $k_u'' = k_b''$.

When working with an intermediate bandpass photometric system, like the uvby system, where the bandpasses are only a few hundred angstrom units wide rather than about 1000Å wide, $k''$ proves to be immeasurably small in each bandpass so that the nasty complication of color-dependent extinction can simply be ignored. It can also be ignored in red and infrared photometry, but here the reason is not the narrowness of the bandpasses but rather the relative flatness of the $k$ versus $\lambda$ curve at longer wavelengths.

Throughout the rest of this book we will often be using the visual bandpass as an illustrative example. Even though, as explained above, $k''$ proves to be immeasurably small in visual bandpasses, we will retain the $k''$ term to keep the illustration general.

## 9.2 Diffraction, Scattering, Dispersion and Seeing

We consider these four effects together because they all act to decollimate the starlight and produce an image which is several arcseconds in size. If they were not occurring, a stellar image would be extremely small. Our sun at a distance of one parsec (slightly closer than $\alpha$ Centauri) would subtend an angle less than 0.01 arcseconds; at a distance of 100 parsecs (a bit closer than the Pleiades) it would subtend an angle less than 0.0001 arcseconds.

To put things in proper perspective, we mention here that for small telescopes with less than ideal tracking and mounting a diaphragm one or two arc *minutes* in diameter might be required to keep the star from approaching its edge too closely. In this case, these effects of seeing, diffraction, scattering, and dispersion will be of negligible concern. For telescopes which can and do use reasonably small diaphragms, these effects do need to be considered.

Even if there were no atmosphere, a point light source observed through a telescope whose objective is of finite diameter would not appear as a point source. Diffraction would cause the image to appear as a round disk of light, called Airy's disk, surrounded by concentric rings of light each less bright than the one closer in and each separated by circles of zero intensity. The angular size of this diffraction pattern, effectively the radius of Airy's disk, is given by

$$\theta = 1.220 \frac{\lambda}{D}, \tag{9.2.1}$$

where $\theta$ is the angular distance in radians from the center (of Airy's disk) to the middle of the first dark ring, D is the diameter of the telescope objective, and $\lambda$ is the wavelength of the light. Both $\lambda$ and D must have the same units. To convert $\theta$ in radians to $\theta$ in arcseconds, simply multiply by 206265. Here you can see that the diffraction pattern is larger for telescopes of smaller aperture. For example, in visible light ($\lambda = 5500$ Å) an 8–inch telescope gives $\theta = 0.68$ arcseconds. Since broadband photometry is not monochromatic, the idealized shape of the diffraction pattern will be smeared out considerably. Young (1974, Eq. 2.1.2) gives a formula for estimating how large a diaphragm you must use to include a certain percentage of the total light in the entire stellar image. A

16–inch telescope with a 40% obscuration by the secondary mirror and used at 5000Å would need a diaphragm 17 arcseconds in diameter to include 99% of the light.

Scattering of light off dust particles on the telescope objective or pinhole defects in the aluminized mirror surface can additionally decollimate the image of a point source, spreading its light even farther from the optical center than diffraction alone would do.

Dispersion of starlight by the earth's atmosphere is a phenomenon which few photometric observers consider but which can be significant at very large zenith distances. Dispersion comes about as the earth's atmosphere refracts light of different wavelengths by different amounts, more so for shorter wavelengths. Young (1974) gives a formula to estimate how important this effect can be. For example, at two airmasses light in the U bandpass of the UBV system will appear 3-arcseconds closer to the zenith than light in the V bandpass which the observer instinctively uses to center a star visually in his diaphragm. Such a 3 arcsecond shift will cause an additional 1% loss of starlight through a diaphragm 20 arcseconds in diameter if a 16–inch telescope is used.

Seeing is the term used to describe the effect of atmospheric turbulence. Many references to this complex phenomenon are given by Young (1974). At one extreme, usually with telescopes of small aperture and with turbulent elements not far above the ground, you see a small stellar image moving rapidly around in the focal plane. At the other extreme, usually with telescopes of large aperture and with turbulent elements far above the ground, there are so many different, very small images (called speckles) moving so rapidly that the effect you see is a blurred image (called the seeing disk). The first extreme degenerates into the second extreme when you make a long exposure measurement, as is the case not only in photography but also in photoelectric photometry. At the very best observing sites, the seeing disk can be somewhat smaller than 1 arcsecond in diameter. Under very poor seeing conditions, which ironically often occur when transparency conditions are best, the seeing disk can be well over 10 arcseconds in diameter. The distribution of brightness within the seeing disk is roughly Gaussian but not exactly.

The overall distribution of brightness within a stellar image decollimated by all of the four effects discussed above (diffraction, dispersion, scattering, and seeing) is very difficult to describe and cannot be represented by a tidy mathematical formula. Nevertheless, it is important for the observer to understand the following points.

1. Equation (9.1.8) tells us that telescopes with apertures smaller than about 2 inches in diameter will have an Airy disk larger than about 2 or 3 arcseconds in diameter, which is the size of typical seeing disks. Small telescopes, therefore, will be diffraction-limited whereas large telescopes will not be.

2. At any one time on a given night the shape of the image's overall brightness distribution should be, on the average, identical for stars of all brightness. For example, the diameter of a diaphragm which would include half of the star image's total light (the image's halfwidth) will be the same for all stars no matter whether they are very faint or very bright. The observer may think a bright star has a larger image and hence needs a larger diaphragm, but that is not so. The image

of a bright star, because it is bright, can be seen out to where its brightness has dropped to 1%, for example, whereas the image of a star ten times fainter can be seen only out to where its brightness has dropped to 10%, thus appearing smaller to the eye. This fact explains why you should *always use the same diaphragm* whenever intercomparing two stars (in differential photometry) or many stars (in all-sky photometry). In principle some starlight must be lost whenever you use a diaphragm of some reasonable size; this lost light is called spillage. Say, for example, a 20-arcsecond diaphragm includes 98% of a star's light on a certain night. Your deflections will be proportional to 98% of the variable's light and 98% of the comparison star's light. Let us call $d(\text{var})$ and $d(\text{comp})$ the deflections you would have measured if you could have included 100% of each star's light. The differential magnitude you compute from your "98% deflections" would be

$$\Delta m = -2^{\text{m}}5 \log_{10} \frac{0.98 d(\text{var})}{0.98 d(\text{comp})} = -2^{\text{m}}5 \log_{10} \frac{d(\text{var})}{d(\text{comp})}, \qquad (9.2.2)$$

which you can see is correct.

**3.** Because atmospheric turbulence is such a random phenomenon and the turbulent elements move so rapidly, the diameter of the seeing disk will fluctuate wildly on a short timescale. Therefore, you can obtain reasonably steady deflections only if you use a diaphragm large enough to include almost all of the light, more than about 97%. Another reason for using a large diaphragm is because dispersion increases strongly as zenith distance increases. If your diaphragm is too small, dispersion might cause the short wavelength bandpass measures taken at large airmasses to be systematically too faint. An additional reason for using a large diaphragm, though not related to atmosphere effects, is to keep a star reasonably centered despite an imperfect telescope drive or mounting.

## 9.3   Scintillation

Turbulence in the earth's atmosphere, in addition to decollimating the stellar image and thereby degrading a nearly point light source into a blurred seeing disk, also causes the total brightness of the image to fluctuate very rapidly, on timescales around 0.1 seconds. Turbulent cells in the upper atmosphere of varying density move across the line of sight at roughly 1 meter per second. These cells act like very weak lenses to spread out and condense the starlight received by a telescope. This random brightness fluctuation is called scintillation (or seeing noise) and essentially is equivalent to what we know as twinkling. The formula given by Young (1974, Eq. 2.1.6.) for the error which scintillation produces in a photoelectric measurement shows that the error is greater at larger zenith distances and larger bandwidths but is smaller for larger telescope apertures and at higher altitudes above sea level. Scintillation will be worse for telescopes with apertures the same size or smaller than the turbulence cells (around 5 inches) and will be less important for telescopes large enough to average the effects of many such cells. For a 16–inch telescope at sea level the error in a 10–second integration ranges from about $\pm\ 0^{\text{m}}001$ at X = 1 to about $\pm\ 0^{\text{m}}01$ at X = 3.

## 9.4   Sky Brightness

The term sky brightness at least is self-explanatory. The amount of sky brightness entering the diaphragm along with the star you are measuring can range from so small as to be insignificant relative to the star being measured to so large as to swamp the star's own brightness. Even the darkest sky has some brightness, mostly as a result of airglow. More often, though, you will be bothered by a sky made bright by moonlight and/or city lights and/or twilight, all of which are exacerbated by water droplets or dust particles in the air. As with so much else in stellar photometry, sky brightness also increases with increasing zenith distance. Your telescope itself can be making the sky background brighter than it otherwise need be. To avoid this, you should shield an open tube telescope from stray lights in the dome, carefully baffle a Cassegrain reflector in the optimum way, and keep your optics as clean and as dust-free as possible.

# Chapter 10

# STANDARD PHOTOMETRIC SYSTEMS

## 10.1 Introduction

The idea of a standard photometric system is an important one in astronomy. This has become especially so with the advent of photoelectric photometry. There are, by now, a very large number of photometric systems (one feels, sometimes, too many) and very much written about them, to say nothing of the huge amount of material published based on these systems.

Perhaps the most valuable aspect of a standard photometric system is that calibrations can be set up once and for all to relate magnitudes and color indices on that system with numerous physical properties of stars. Depending on the particular photometric system, such properties can include temperature, absolute luminosity, surface gravity, degree of absorption by interstellar material, chemical composition, and age. Thus, any one astronomer can observe a certain star or group of stars on that photometric system and immediately draw on those calibrations to learn much about his stars, without having to duplicate the large amount of work involved in setting up those calibrations.

Another value of standard photometric systems, especially for variable star photometry, is that observations of the same star by many different observers with quite different photometric equipment is of value. The data can be combined on the same light curve to reveal the nature of the variability in greater detail or with fewer gaps than possibly could be done by any one observer.

Useful references pertaining to standard photometric systems in general include Fernie (1983b), Bessell (1979), Golay (1974), Crawford (1969), Johnson, Mitchell, Iriarte, and Wisniewski (1966), Kron and Smith (1951), and Kron, White, and Gascoigne (1953).

While refraining from a comprehensive summary or review of photometric systems, we would like to say that they can be divided into three types, depending on the width of the bandpasses: wide-band, intermediate-band, and narrow-band. Wide-band photometric systems use bandpasses around 20% of the wavelength in question. Thus, for blue and visual this would be around 1000Å. The best known example, and the most widely used, is of course the UBV system, sometimes called the Johnson system. Such a system can look only at gross features in the continuum of a stellar spectrum, such as the way its overall slope changes with

161

temperature or the way the Balmer jump is influenced by temperature and surface gravity. Intermediate-band photometry uses bandpasses only a few hundred angstrom units wide. The best known example is the uvby system, also called the Strömgren four-color system. Such a system can look at somewhat finer (as well as gross) features in the spectrum. Narrow-band photometry uses bandpasses only a few tens of angstrom units wide. A good example of this is the $\beta$ system, which measures the amount of absorption in the H$\beta$ line at 4861Å.

Clearly, measures at many narrow wavelength intervals are capable of yielding more detailed information about the physical conditions in a stellar atmosphere, but at the expense of having to "throw away" much of the light. Thus, a telescope which could conveniently do V–band photometry of a 10th magnitude star could do $\beta$ photometry only for stars brighter than about 6th magnitude. Moreover, intermediate and narrow-band filters are usually interference filters, which are considerably more expensive than glass filters and are much more susceptible to deterioration with time. These are practical reasons why, despite the limitations of wide-band photometry relative to narrow-band, wide-band photometry continues to be the choice of many observers.

A standard photometric system is defined by a list of stars each having a magnitude (or a number of magnitudes, if it is a multifilter photometric system) assigned to it. Although a photometric system is usually set up by means of observations made at one observatory site with a specific telescope, photomultiplier, filters, etc., it is not the equipment which defines the standard system; rather it is the published magnitudes, thus called standard magnitudes, assigned to the certain stars, thus called standard stars.

Magnitudes determined by any observer with a given telescope and photometric equipment are called natural magnitudes or instrumental magnitudes. If that observer wishes to do photometry on some standard photometric system, he uses his own photometric equipment to determine extraterrestrial instrumental magnitudes of those same standard stars and then attempts to find a relation (such as a simple formula), which relates his extraterrestrial instrumental magnitudes to the standard magnitudes. This relation is called the transformation. If a simple relation can be found which, for stars of all types, reproduces the standard magnitudes to whatever accuracy would reasonably be expected, usually something like $0^{m}02$, then the transformation is called a successful one.

To maximize the likelihood of a successful transformation, your own photometric equipment should as closely as practical duplicate each bandpass in the standard system, both in effective wavelength and bandpass width. Many elements collectively determine a bandpass: the reflectivity of the objective mirror and any secondary mirrors or diagonal flats, the transmissivity of the objective lens and Fabry lens and any prisms and windows, the transmission curve of the filters, the relative sensitivity of the photosensitive surface at various wavelengths, and the transmissivity of the atmosphere where the telescope is situated. Using the same type of photomultiplier and filters which were used to set up the standard system originally will not necessarily guarantee a good transformation, because those are not the only relevant factors. As Hardie (1962) explains, it often proves necessary to choose slightly different filters deliberately to offset some other factors in order to achieve a good match. As you can see, therefore, there is no such thing as a set of "standard filters."

## 10.2 Wide Band Systems

The most widely used wide-band standard photometric system is the UBV system. The RIJKLMN photometric system, reaching out to 10 microns in the infrared, is sometimes considered an extension of the UBV system. The two together, UBVRIJKLMN, are sometimes referred to irreverently as the alphabet soup photometric system. The original instrumentation which set up the UBV system used an RCA 1P21 photomultiplier refrigerated with dry ice. As filters, it used Corning No. 9863 of standard optical thickness for U, Corning No. 5030 of standard optical thickness cemented to 2 mm of Schott GG–13 for B, and Corning No. 3384 of standard optical thickness for V. A modified ITT FW 118 photomultiplier, also refrigerated with dry ice, was used for R and I. A PbS photoconductive cell refrigerated with liquid nitrogen was used for IJK. An InSb photovoltaic cell refrigerated with liquid helium was used for JKLMN. The effective wavelengths and effective bandwidths, taken from Allen (1973), are given in Table 10–1. It should be noted that effective wavelength varies somewhat depending on the color of the star being observed; for example, the effective wavelength in U varies between 3500Å for a star of B–V = –0.2 to 3800Å for a star of B–V = +1.2. The response curves, which define the detailed shape of the UBV bandpasses, are given in numerical form by Allen (1973). Allen (1973) also cites most of the references pertinent to the alphabet soup photometric system.

When the UBV system was set up, use was made of 10 so-called primary standard stars, but a much larger list of stars can be considered standard stars. The original such list (Johnson and Morgan 1953) included 290 stars with UBV magnitude determinations. Shortly thereafter, Johnson and Harris (1954) published "three-color observations of 108 stars intended for use as photometric standards." The "list of [94] well-observed standard stars, thoroughly tied in with the ten primary standards," published later by Johnson (1963), appears to be much the same list. Many observers have found the so-called *Arizona-Tonantzintla Catalogue* of Iriarte et al. (1965) very helpful. This contains not only UBV but also RI magnitudes of 1325 bright stars. Since these magnitudes were determined many years after the UBV system originally was set up and relied on different equipment at different observatories, they cannot be considered bona fide standards. As Iriarte et al. (1965) themselves explain, "It should be remembered that the new Catalina-Tonantzintla observations do not define the UBV system." They are magnitudes on the UBV system, however, and using them as standards will not likely cause significant trouble if care is taken to avoid those noted as known or suspected variables and those observed only once. *The Astronomical Almanac* now publishes "a selection of 107 stars to serve as standards of the [Johnson] UBVRI photometric system." Hall (1983d) points out that this newest list of standards promises to be useful by being easily accessible and being composed of conveniently bright stars.

| TABLE 10-1 | | | | | | | |
| --- | --- | --- | --- | --- | --- | --- | --- |
| EFFECTIVE WAVELENGTHS AND BANDWIDTHS | | | | | | | |
| (IN MICRONS) | | | | | | | |
| | $\lambda$ | $\Delta\lambda$ | | $\lambda$ | $\Delta\lambda$ | | $\lambda$ | $\Delta\lambda$ |

| | $\lambda$ | $\Delta\lambda$ | | $\lambda$ | $\Delta\lambda$ | | $\lambda$ | $\Delta\lambda$ |
| --- | --- | --- | --- | --- | --- | --- | --- | --- |
| U | 0.365 | 0.068 | R | 0.70 | 0.22 | u | 0.350 | 0.034 |
| B | 0.440 | 0.098 | I | 0.90 | 0.24 | v | 0.411 | 0.020 |
| V | 0.548 | 0.089 | J | 1.25 | 0.38 | b | 0.467 | 0.016 |
| | | | K | 2.20 | 0.48 | y | 0.547 | 0.024 |
| | | | L | 3.41 | 0.70 | | | |
| | | | M | 5.00 | – | | | |
| | | | N | 10.20 | – | | | |

## 10.3   Intermediate and Narrow Band Systems

The most widely used intermediate-band photometric system is the uvby system. The most widely used narrow-band photometric system is the $\beta$ system. The two, often used together as the uvby/$\beta$ system, are described by Strömgren (1966).

The effective wavelengths and effective bandwidths for the uvby system, taken from Allen (1973), are given in Table 10–1. The letters u, v, b, and y refer to magnitudes in the ultraviolet, violet, blue, and yellow spectral regions, respectively. The bandpasses are narrow enough and were chosen with sufficient foresight that they are defined virtually by the filters alone and hence are not a function of the photocathode's spectral response, the altitude of the observing site, or the color of the star being observed. Moreover, the atmospheric extinction in these bandpasses should not be color-dependent; in other words, you can forget about the $k''$ terms. Transmission curves of representative filters used to set up this photometric system are given by Crawford and Barnes (1970) along with magnitudes of 304 stars which they present as standard stars to define the uvby system. Actually, the uvby system is a system of color indices, not individual magnitudes. These indices are $b-y$, $m_1 = (v-b)-(b-y)$, and $c_1 = (u-v)-(v-b)$. Since the effective wavelength of the y bandpass is nearly identical to that of the V bandpass of the UBV system and because the yellow spectral region in most stars suffers from no gross discontinuities, y can be transformed to V with little or no intrinsic scatter and thus can be regarded as equivalent to V. For somewhat similar reasons, the b–y index can be transformed reasonably well into the B–V index of the UBV system, with a transformation coefficient of approximately $\mu = 1.5$, although be warned that the transformation curve (b–y plotted versus B–V) is not a perfectly straight line.

The $\beta$ system is used to study the strength of the H$\beta$ absorption, which occurs at 4861Å in stars at rest with respect to the observer. It employs two interference filters both with peak transmission as near 4861Å as possible, a narrow one with a halfwidth (FWHM) around 30Å and a wider one with a halfwidth around 150Å. This system makes use of an index, called $\beta$, which is based on the magnitude difference corresponding to the ratio of the light intensity measured through the two filters. Because the two filters have identical effective wavelengths, measured values of $\beta$ are independent of atmospheric extinction and also interstellar absorption, a very attractive convenience. Crawford and Mander (1966) give a list of 80 stars and corresponding $\beta$ magnitudes which they recommend for use as standards to define the $\beta$ system.

## 10.4 Transformation Relations

Let us take the UBV system as an example to illustrate the form which transformation relations usually take:

$$V = v_o + \epsilon_v(B - V) + Z_v, \qquad (10.4.1)$$

$$B - V = \mu(b - v)_o + Z_{b-v}, \qquad (10.4.2)$$

and

$$U - B = \psi(u - b)_o + Z_{u-b}. \qquad (10.4.3)$$

Here, V, B, and U represent the standard magnitudes in the three bandpasses; $v$, $b$, and $u$ represent the corresponding magnitudes in the instrumental system, and the subscript labels them as outside-the-atmosphere or extraterrestrial. Notice that $(b - v)_o$ here is just shorthand for $b_o - v_o$. It has been customary, though admittedly a bit awkward in some situations, to deal with one magnitude (V) and two color indices (B–V and U–B) rather than the three individual magnitudes (V, B, and U) separately. The three transformation coefficients are $\epsilon_v$, $\mu$ , and $\psi$ whereas the last term in each equation is the zero point. It might be that the extraterrestrial instrumental magnitudes cannot be related to the standard magnitudes with simple linear equations like the ones above, in which case one might have to resort to a more complex equation or perhaps even to a graphical relation. Another possibility to beware of is that different types of stars (dwarfs versus giants, reddened O–type stars versus unreddened M–type stars, stars of peculiar versus normal chemical composition, etc.) might give different relations. Nonlinear and multivalued transformations are most likely to occur if your instrumental system is not closely matched to the standard system.

Some astronomers prefer to work with the individual magnitudes throughout, rather than with the color indices. The corresponding three equations then would be

$$V = v_o + \epsilon_v(B - V) + Z_v, \qquad (10.4.4)$$

$$B = b_o + \epsilon_b(B - V) + Z_b, \qquad (10.4.5)$$

and

$$U = u_o + \epsilon_u(U - B) + Z_u, \qquad (10.4.6)$$

where obviously Eq. (10.4.4) is identical to Eq. (10.4.1).

Let us see what the values of the transformation coefficients can tell us. If your instrumental system is matched so closely to the standard system that your extraterrestrial magnitudes need no transformation at all except for the additive zero points, then you would have $\epsilon_v = \epsilon_b = \epsilon_u = 0$ and $\mu = \psi = 1$. If any coefficient differs more than about 0.1 from these ideal values, that indicates that the match is not very close and that the transformation might not be linear or singlevalued. If $\epsilon_v$ is positive (or negative), then the v bandpass lies to the red (or blue) of the V bandpass. The situation is analogous for $\epsilon_b$ and $\epsilon_u$. If $\mu$ is greater (or less) than 1, then the $v$ and $b$ bandpasses are closer together (or farther apart) than are the V and B bandpasses. The situation is analagous for $\psi$.

Many factors go into determining the zero points, especially those for single magnitudes: $Z_v$, $Z_b$, and $Z_u$. These factors include:

1. The telescope aperture,

2. The photomultiplier's sensitivity (a combination of the quantum efficiency of the photocathode and the internal amplification produced by the successive stages),

3. The filter halfwidths and peak transmissions,

4. The amplification factor produced by the amplifier or the interval of time over which a photon counting device sums the incoming pulses,

5. The sensitivity of the chart recorder, and

6. The scale used to measure the deflections on the chart paper.

Meaningful photometry depends on keeping the zero point constant throughout any one night of photometry, although differential photometry, discussed in the next chapter, requires only that the zero point remain constant for shorter intervals during a night. It may help, therefore, to list some of the actions which *will change* the zero point:

1. Washing or realuminizing the mirror,

2. Getting a grease smudge on a filter or some other optical element,

3. Changing amplifiers, raising or lowering the voltage supplied to the photomultiplier, changing gain on the strip chart recorder,

4. Reorienting the photomultiplier tube,

5. Exposing the photocathode to a strong light, and

6. Switching the units used to measure the deflections on the strip chart paper.

It might be asked why no coefficient appears in front of $v_o$, $b_o$, $u_o$. That is because their coefficients must be exactly 1. In other words, if you plotted V versus $v_o$ for standards which had exactly the same value of B–V, the slope of the resulting line must be exactly 1 ($45°$). If it were not, then some element of your photometric system (the photomultiplier, the amplifier, the recorder) must not be responding linearly (in direct proportion) to incoming light. The trouble could be saturation, leaking, gassing in the photomultiplier; malfunctioning or incorrect calibration of the amplifier; or failure to account for coincident pulses going uncounted by the pulse counter. With equipment malfunctioning in this way, you simply cannot do meaningful photometry.

Although the UBV system has been used in this chapter and the next to illustrate the structure, significance, and application of equations which correct for atmospheric extinction and transform from one's instrumental system to a standard system, the reader should be able to generalize to other photometric systems. More often than not, ironically, the job is easier than with the UBV system, not harder. We have mentioned already that $k''$ can be ignored in intermediate and narrow band photometry and that $k'$ is smaller in R and I than in U, B, or V. Just in the last couple of years, thanks to the widespread use of good but reasonably priced solid-state photometers such as the SSP–3 of Optec (Sanders and Persha 1983), R and I photometry is becoming popular. An entire issue of the *I.A.P.P.P. Communications* (No. 14) was devoted to this subject.

# Chapter 11

# DIFFERENTIAL PHOTOMETRY

## 11.1 Background

In this book we explain how to do differential photometry, the technique which yields most accurate results when doing photometry of some one star, usually a variable or suspected variable. This technique is most tolerant of skies of less-than-ideal transparency, and we recommend it as most suitable for amateur and small college observatories.

All-sky photometry results in the determination of standard magnitudes of a number of stars, called program stars, usually distributed all over the sky. It involves observing the program stars and a number of standard stars together on the same night, determining the principal extinction coefficients accurately for that night, making sure that the zero point in each bandpass of the instrumental photometric system remains constant or drifts only slightly or gradually throughout that night, and knowing the transformation coefficients which describe the instrumental system at that time. This is discussed in somewhat more detail by Hardie (1962). Needless to say, the sky must be of high photometric quality, i.e., the atmospheric transparency must be uniform over the entire sky and must remain constant throughout the entire night. These requirements are sufficiently stringent and the overall procedure sufficiently complex that all-sky photometry is best left for the competent professional research astronomer to do with excellent photometric equipment at very good observing sites, i.e., ones at high altitude with dark skies and a large number of photometric nights each year.

## 11.2 Advantages of Differential Photometry

Differential photometry involves determining the magnitude difference between one star (the variable) and another star (the comparison star) very close to it in the sky. The fundamental advantages of differential photometry over all-sky photometry are many. You can use a night which is clear for only a short time and in principle even one during which only a portion of the sky is clear. Moreover, the comparison star is selected so that, when determining the magnitude differences, most of the crucial effects which all-sky photometry must deal with simply "cancel out."

1. If the variable and its comparison are close together in the sky, then atmospheric extinction affects both by about the same amount and only a quite small correc-

tion is needed to correct for the small differential atmospheric extinction.

2. If the variable and its comparison have very nearly the same color, then the difference in wavelength response between the instrumental system and the standard system affects both by about the same amount and only a quite small correction is needed to correct for the small differential transformation.

3. If the variable and its comparison are fairly similar in brightness, then they can be measured with the same coarse gain setting on the amplifier and you are not vulnerable to any inaccuracy in determining the calibration of the coarse gain resistors or to the variation they may undergo with temperature changes.

4. Because the variable and its comparison are compared within a few minutes, the zero point of your photometric system need hold constant only for those few minutes. Actually, because the comparison star is measured immediately before and after the variable, and its *average* brightness is used for the comparison, the zero point may even drift slightly provided it does so gradually rather than abruptly.

5. Finally, as with the zero point, the sky transparency need remain constant or change only gradually during those few minutes required to bracket the variable's deflection between two comparison star deflections.

Individual magnitudes determined by all-sky photometry are almost never more accurate than $0^m01$, $0^m02$, or $0^m03$, even when determined under the best conditions at the best observatories by the best photometrists. Individual differential magnitude determinations, however, typically can be accurate to within about $\pm$ $0^m01$, even when made by amateurs with small telescopes under somewhat adverse observing conditions, and often are better, around $\pm$ $0^m007$. Two amateurs with small telescopes achieved an accuracy of $\pm$ $0^m003$ (Landis, Louth, Hall 1985). It is of little practical importance that the absolute brightness level of a light curve based on differential photometry is only as certain as the absolute brightness of the comparison star. This is because the variable's light *changes* are of interest. Moreover, different series of differential magnitudes (from different observers, from different years, etc.) can be combined or inter-compared without sacrificing the accuracy inherent in differential photometry provided the same comparison star is used for both series.

## 11.3   The Equations

The equations which describe the effect of atmospheric extinction and transformation on differential photometry follow by simply writing the corresponding equations down twice, once for the variable and once for its comparison, subtracting, and using the symbol $\Delta$ to indicate difference in the sense variable minus comparison. Let us use the UBV system as an example of a wide-band three-filter photometric system.

In differential photometry we deal with the (usually very small) difference in $X$ between the variable and its comparison, calling it the differential airmass $\Delta X$. X has to be computed with considerable numerical accuracy to determine the very small difference in $X$ between two stars which are very close together in the sky. Therefore, it is helpful sometimes to compute $\Delta X$ directly with the formula derived by Hardie (1962, §5).

The result for differential atmospheric extinction, derived from equations like Eq. (9.1.7), is

$$\Delta v_o = \Delta v - k_v' \Delta X - k_v'' \bar{X} \Delta(B-V), \tag{11.3.1}$$

$$\Delta b_o = \Delta b - k_b' \Delta X - k_b'' \bar{X} \Delta(B-V), \tag{11.3.2}$$

and

$$\Delta u_o = \Delta u - k_u' \Delta X - k_u'' \bar{X} \Delta(U-B), \tag{11.3.3}$$

where $\bar{X}$ is the average of $X$ for the variable and $X$ for its comparison star.

The result for differential transformation, derived from Eqs. (10.4.1), (10.4.2), and (10.4.3), is

$$\Delta V = \Delta v_o + \epsilon_v \Delta(B-V), \tag{11.3.4}$$

$$\Delta(B-V) = \mu \Delta(b-v)_o, \tag{11.3.5}$$

and

$$\Delta(U-B) = \psi \Delta(u-b)_o. \tag{11.3.6}$$

Or, derived from the alternate Eqs. (10.4.4), (10.4.5), and (10.4.6), it is

$$\Delta V = \Delta v_o + \epsilon_v \Delta(B-V), \tag{11.3.7}$$

$$\Delta B = \Delta b_o + \epsilon_b \Delta(B-V), \tag{11.3.8}$$

and

$$\Delta U = \Delta u_o + \epsilon_u \Delta(U-B). \tag{11.3.9}$$

Notice that all of the zero points have dropped out when we do differential photometry.

At this point, we can conveniently show how $\epsilon_b$ is related to $\mu$. The difference between Eqs. (11.2.8) and (11.2.7) gives us $\Delta(B-V)$, which can be set equal to Eq. (11.2.5). Then, recalling that $\Delta b_o - \Delta v_o = \Delta(b-v)_o$, it follows very quickly that

$$\frac{1}{\mu} = 1 - \epsilon_b + \epsilon_v. \tag{11.3.10}$$

Unfortunately, it is much more difficult to relate $\epsilon_u$ to $\psi$, because $\epsilon_u$ and $\psi$ are coefficients of $\Delta(U-B)$, not $\Delta(B-V)$.

# Chapter 12

# OBSERVING TECHNIQUES

## 12.1   Choosing a Comparison Star

One requirement for the comparison star, clearly the most crucial, is so obvious it almost does not need stating but yet sometimes is overlooked, with disasterous consequences. The comparison star must not, itself, be a variable. If you are participating in a cooperative campaign or if the variable has been observed before by others, it may be that a certain comparison star has been used before and found to have been constant. If a search through the literature shows that you are the first to be observing a certain variable photoelectrically, then as a first precaution you can check in the *General Catalogue of Variable Stars* that the comparison star tentatively selected is not a known variable. A special *Supplement to the Third* (1969–70) *Edition* is a list of all known and suspected variables ordered according to 1950 right ascension, ideal for this purpose. It is advisable, nevertheless, to select a check star, a second or auxiliary comparison star. Typical procedure is to observe the check star differentially with respect to the comparison once each night. Notice, however, that this procedure at best can only verify that the comparison star was not noticeably variable. If the check versus comparison differential magnitudes do show a variation, you cannot immediately know which of the two was varying. Moreover, if the comparison star was varying, it is likely that the entire series of variable star differential magnitudes cannot be salvaged and must be discarded. The extemely careful observer can bracket the variable each time between two different comparison stars and also measure a check star. But the law of diminishing returns can make that much care counterproductive, leaving little time to observe the variable itself.

Perhaps the next most important requirement is that the variable and its comparison should be as close together in the sky as possible. The obvious reason for this is to minimize the value of $\Delta X$ which occurs in Eqs. (11.3.1), (11.3.2), and (11.3.3). As a rule of thumb, try to avoid comparison stars farther than about a degree away. Admittedly, this is not always possible when working with bright stars because there simply are not that many bright stars in the sky. The angular separation between the variable and its comparison would be *too small*, however, if an inconveniently small diaphragm was required to avoid light contamination from one when observing the other. In differential photometry, it is important to make the intercomparison between the variable and its comparison as quickly as possible. Thus, it is an additional advantage if the comparison star can be selected such that the telescope need move in only one coordinate (for example, declination) to go from one star to the other or if the two are sufficiently close that both are visible together in the finder, or better still, in the previewing eyepiece.

It can be seen in Eqs. (11.3.7), (11.3.8), and (11.3.9) that differential transformation involves adding a correction which is proportional to the difference in color index between the variable and its comparison. Thus, even though you should have determined your transformation coefficients accurately, the accuracy of the transformation process is greater to the extent that the variable and its comparison have nearly the same color. Similarly, a good color match minimizes errors introduced in the $k''$ term of Eqs. (11.3.1), (11.3.2), and (11.3.3). A color index difference of only 0ᵐ1 is the best you can reasonably hope for; 0ᵐ5 would be tolerable; and a difference as large as 1ᵐ0 should be avoided.

If possible, the variable and its comparison star should be close enough in brightness that both can be observed with the same coarse gain setting on the amplifier. Thus, any inaccuracies in the calibration of the large resistors will not be a factor. Although those large resistors can and should be calibrated accurately, this is not always possible because their resistance often varies appreciably as a function of temperature, which will change from night to night and during a given night. An amplifier which can cover a large range with its fine gain settings (like the Oliver amplifier and the Stokes amplifier, which go from 0.0 to 3.5 with their half-magnitude steps) makes it more likely that the same coarse gain step can be used for both stars. Of course, if you use pulse counting (and to some extent if you use a digital voltmeter), then the problem with different coarse gain settings does not arise. The criterion of comparable brightness is important for other reasons when dealing with very bright stars, however. Very bright stars can cause fatigue or saturation in a photomultiplier and can lead to a large number of coincident pulses going uncounted by pulse counters, more than can be accounted for accurately by the coincident pulse correction discussed later. The more nearly equal in brightness two very bright stars are, however, the more nearly those two effects will cancel out when the differential magnitudes are determined. When working with a faint variable, a general rule is that the comparison star should, if possible, be bright enough not to be photon-noise-limited. Although you cannot make the variable brighter, you are free to try to select a comparison star which is bright enough. The reader will see that this last rule (don't choose a comparison star too faint) can conflict with the first rule (have the variable and comparison comparable in brightness), in which case a compromise should be made.

## 12.2   Monitoring the Sky

A useful prelude to observing is to monitor an arbitrary star to ascertain the photometric quality of the sky. This is best done by selecting a relatively bright star rather near the zenith, centering it in an overly large diaphragm, turning up the voltage to its normal operating level, and adjusting the amplifier to produce a near full-scale deflection. If all is well, 10 or 20 minutes should produce a nearly horizontal, though somewhat ragged, tracing on your strip chart recorder. The raggedness, sometimes referred to as grass, is the result of photon noise and/or scintillation; but if the tracing is not nearly horizontal, then something is wrong.

If the star has drifted out of the diaphragm (and it should not drift out of an overly large diaphragm in just 10 or 15 minutes), then perhaps the telescope needs to be balanced more carefully or the drive rate adjusted. If the tracing is going steadily down, perhaps the sky is not yet entirely dark but soon will be, something which could be verified by comparing sky deflections taken before and after this monitoring deflection. A changing level could mean that either the photomultiplier or the amplifier or the power supply has not yet adapted to the recently applied voltage. After a half hour or so, though, they all should be. Of course, if you follow the practice some photometrists do of leaving the power supply and the high voltage and the amplifier on 24 hours a day, or turning them on early in the evening, this last possibility should not apply.

With all of the above factors excluded, a variable tracing would probably indicate that one problem which no photometrist has the power to solve: variable atmospheric transparency, commonly known as clouds. Of course, clouds can and should be looked for in the sky by eye, but they are often difficult or impossible to see on dark moonless nights. If the tracing is changing levels very gradually with time and is covering a range less than about 5% during a 20–minute interval, then it may be possible to obtain usable differential photometry if the variable and its comparison are intercompared relatively quickly. However, such a night is at the threshold of being deemed non-photometric. Certainly, if the range is greater than about 5% or if the fluctuations are at all erratic rather than gradual, then usable photometry on that night should not be anticipated.

## 12.3 How to Make a Deflection

The word deflection is used to describe a single measurement of the brightness of a star (or the sky) with your photometric equipment. Originally, that word referred to a deflection of the light beam or the needle on a milliammeter but it has been generalized now to refer to the displacement of the pen on a strip chart recorder or the number of photons counted with pulse counting equipment or the number of counts registered with a digital recording device.

The amplifier gain should be selected so as to make the deflection as nearly full scale as possible without "pinning," i.e., overshooting full scale. If your amplifier has half-magnitude (1.585×) gain steps, deflections less than about $\frac{2}{3}$ of full scale can be avoided. If your amplifier has a 2× gain step, deflections less than about $\frac{1}{2}$ full scale can be avoided. These considerations do not apply, however, if you are using digital recording devices with a large dynamic range.

The proper procedure involves making a deflection on the star followed immediately by a deflection on the sky nearby, with nothing changed except the direction the telescope is pointed (or, for those telescopes so equipped, the diaphragm offset). You should not change the voltage on the photomultiplier, the gain setting on the amplifier, the diaphragm being used, any time constant, any zero point adjustment, or the filter. The reason for this is clear. The star + sky deflection gives you a number which represents the star's brightness + the sky brightness + the dark current + any zero point which might be introduced by the electronics of the amplifier or the chart recorder. The sky deflection gives you a number which represents the sky brightness + the same dark current +

the same zero point.  Therefore, simple subtraction (star + sky deflection – sky deflection) gives you the net star deflection, which must necessarily represent the star's brightness only.  The length or size of this net deflection will be directly proportional to the star's brightness provided the response of the photomultiplier, the amplifier, and the chart recorder are all linear, as they all must be if operating properly.

Often an astronomer observing bright stars notices that the sky brightness is so small as to be essentially immeasurable and hence feels he does not need the sky deflection, relying on the dark current or zero level established earlier that night. This shortcut is dangerous for several reasons and should not be taken.  Although the zero point, dark current, and sky brightness very rarely change appreciably within 10 minutes or so, they can change appreciably within a couple of hours or so, for a variety of reasons.  First, although the amplifier should be balanced at the beginning of the night, i.e., adjusted so that the fine gain steps do not alter the zero point, balancing can never be more than approximate.  Moreover, the amplifier very often will get out of balance as the night progresses, because of temperature changes for example.  Thus, the zero point at one fine gain setting cannot be relied upon to be the same as the zero point on another fine gain setting or even on the same fine gain setting a couple of hours before.  Second, the dark current itself will change gradually as the temperature changes throughout the night, especially if the photomultiplier is not refrigerated.  Third, although the actual sky brightness may be immeasurably small at one time, it can increase later in the night if, for example, the moon rises or the amount of haze in the atmosphere increases.

If you have a microscope postviewer and if you do not use an automatic sky offset, you should look in immediately after the star deflection, see if the star is still centered in the diaphragm, and then watch it move out of the diaphragm as the telescope is offset for the sky deflection.  If the star was found too close to the diaphragm edge, then that deflection should be deleted and taken again.

A deflection on the star or the sky should involve exposing the photocell to its light for not more than about 30 seconds.  Because of the oscillations in the milliammeter needle or the grass atop the chart recorder's deflection or the fluctuating count rate with digital recording equipment, the observer is tempted to "sit on the star" for a long time in order to obtain a reliable reading; but sitting on it too long can be counterproductive.  Changes in sky transparency generally prove to be the greatest source of error, and that error is minimized most effectively by bracketing the variable's deflection between two comparison star deflections as quickly as possible and by obtaining as many *different* bracketings as time permits.

To obtain a ∼ 30 second deflection you have several options.  With a chart recorder or milliammeter, for example, you can simply sit on the star for 30 seconds.  If you have an automatic star-to-sky switch, you can make two 15–second deflections with a sky deflection between.  If you have a charge-integrating or count-timing device, you can make three 10–second readings in rapid succession. Notice that 30 seconds is pretty much an upper limit, unless the star is extremely faint.  If the program stars are sufficiently bright that meaningful measures can be obtained in less time, you can do with 20 or even 10 seconds and thus move faster between the variable and its comparison.  Those doing bonafide pulse counting (i.e., not digital voltmeter) can be guided by the following rule: accumulate pulse counts at least until a total of 10,000 is achieved. This follows because n = 10,000 is the number for which, by the "$\sqrt{n}$ rule", photon noise will amount to an uncertainty of $\pm$ 0$^{\mathrm{m}}$01.

You should always scrutinize your tracing or count rates critically, as they are being made, for telltale signs of trouble. Examples could be passing patches of fog, the star drifting out of the diaphragm, the star drifting back into the diaphragm during the sky offset, sluggish response with too slow a time constant, erratic response as strain is exerted on loose electrical connections, etc.

## 12.4 Recommended Basic Sequence for Differential Photometry

We would like to recommend a specific sequence of deflections with one filter suitable for determining the brightness of a variable which does not change very much during the course of one night. This is shown in Fig. 12–1 as a sample strip chart recording, although the essential features apply also to readings on a milliammeter or a digital recording device. The main reason for bracketing the variable *three* times is to deal with the problem of variable sky transparency. Moreover, not only do you get three estimates of the differential magnitude instead of just one, you can examine the extent to which the three are consistent with each other and use simple statistics, discussed in Chapter 14, to derive an estimate of the uncertainty of the average. The value of any scientific measurement is increased significantly if it comes with a quantitative estimate of its reliability.

## 12.5 Possible Reasons for Altering the Basic Sequence

The sequence recommended in the preceding section and shown in Fig. 12–1 can and should be altered in certain circumstances, and the observer should understand clearly the motivation for any alteration. The choice of a particular sequence should not be merely a matter of personal preference or a whim. It should be chosen specifically to minimize error and maximize accuracy given the observing situation forced on you at that time.

The check star deflection can be omitted if it is already known with some confidence that the comparison star is not variable, or if the observer is participating in a cooperative campaign and some other observers are responsible for verifying its constancy. It may also be omitted if for some reason the observer must obtain an observation of the variable very quickly and simply has no time for the check star.

If the variable is varying rapidly minute by minute (an eclipsing binary like RZ Cassiopeiae with a deep minimum of brief duration, an occultation event, etc.), you should obtain more than one variable star deflection between comparison star deflections. Or if the variable is being observed very near the zenith and the sky transparency on that particular night is remaining extremely constant, you can take relatively fewer comparison star deflections.

If the variable and its comparison star are so similar in brightness that both are being observed on the very same amplifier gain setting, coarse *and* fine, then

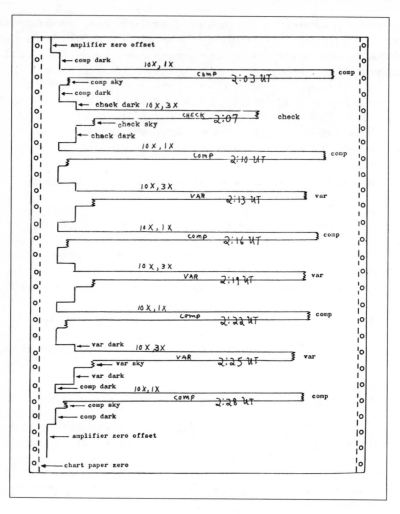

**Figure 12–1.**  *Sample strip chart recording from differential photometry.*

you can take fewer sky deflections, since in this case they *will* be the same for both stars. This same shortcut is appropriate if you are using pulse counting or some other type of digital output which has a large dynamic range, in which case stars of different brightness usually can be measured with the same gain setting.

## 12.6   Multifilter Differential Photometry

There are many good reasons why single filter photometry is sufficient for a given observing program, as explained in Chapter 15. You should, in fact, do single filter photometry whenever it is sufficient, because it is capable of greater accuracy than multifilter photometry, if only because no time is lost changing filters and because the bracketing of the variable between two comparison star deflections can be accomplished more rapidly. Nevertheless, there are occasions when multifilter photometry is needed.

For an observer who moves his telescope manually for the sky offset a good sequence for one multifilter deflection of the variable would be

$$V_v V_b V_u S_v S_b S_u,$$

where V represents a 30–second deflection on the variable and S represents a 30–second deflection on the sky. The sequence would be analagous for the comparison star. The crucial aspect to note is that, when the telescope is moved for the sky offset, the sky deflections with each filter must be made with the same gain setting which was used for the corresponding star deflection.

For an observatory which has a manual or automatic star-to-sky switch, in which case the telescope itself does not have to be moved for the sky offset, or an automated stepper motor, in which case the telescope can be moved quickly and automatically for the sky offset, a good sequence for one multifilter deflection of the variable would be

$$V_v S_v V_v V_b S_b V_b V_u S_u V_u,$$

where now V represents one 15–second deflection on the variable and S represents a 30–second deflection on the sky. Again, the sequence would be analagous for the comparison star.

If you are doing multifilter photometry of a rather slow variable, it is acceptable to consider the deflections with all filters to have been made at the same time, since the variable will not have changed much in brightness during the minute or so between filter changes. This shortcut is a great convenience when recording and later analyzing the data. If the light is changing very rapidly, however, such as during an eclipse of RZ Cas, you must record the time of each filter's deflection separately. A convenient alternative with such rapid variables would be a symmetrical multifilter sequence, for example,

$$V_v V_b V_u V_u V_b V_v S_v S_b S_u,$$

where V represents a 15–second deflection on the variable and S represents a 30–second deflection on the sky. The average of the two V deflections, of the two B deflections, and of the two U deflections would be equivalent to 30–second deflections in each filter taken at the same instant, i.e., at the midpoint of the sequence. The three sky deflections could be taken at the end, as shown above, or at the beginning; it does not matter. Of course, a symmetrical sequence like this would not be needed for the comparison star because only the variable is varying rapidly, not the comparison star.

If the variable is a slow one, as many are, not changing much during the course of one night and if sky transparency conditions require the observer to move as quickly as possible between variable and comparison, then you can and should use the basic sequence described in Section 12.4, doing so separately for each of the filters.

## 12.7   Choice of Diaphragm

The first requirement about choice of diaphragm is that, no matter what size
you use, *use the same one* when intercomparing the variable and its comparison.
The reason for this was explained in Chapter 9. A switch to another diaphragm
may be made midway through a night for a variety of reasons, but the bracketing
of a variable star deflection between two comparison star deflections must be
completed with the same diaphragm. After the switch is made and the sequence
continued, the next deflection must be again on the comparison star.

As a general rule, you should use as large a diaphragm as possible. This
minimizes the danger that the star will drift too close to the edge of or out of
the diaphragm while the deflection is being made, the error produced by erratic
fluctuations in seeing disk size, and any systematic error due to dispersion. There
are basically only two reasons why you would resort to a smaller diaphragm.
First, a smaller diaphragm may be needed to exclude a nearby visual companion.
The general rule is "in or out but not on the edge." For example, if the visual
companion in question is known to be, say, 30 arcseconds away, then a diaphragm
30 arcseconds in *diameter* will be adequate for excluding it; but a diaphragm 60
arcseconds in diameter would be the worst possible choice. That would place
the companion right on the edge of the diaphragm, in which case it would be
sometimes in and sometimes out, depending on how precisely the variable was
centered and how evenly the telescope was tracking, and the observer could never
know which. Second, you should go to a smaller diaphragm if the sky is too bright
relative to the brightness of the star. It is OK if the sky deflection is large enough
to be seen; only if it is too large a fraction of the star's own light will it appreciably
affect the signal-to-noise ratio of the net star deflection. As a rule of thumb, go to
a smaller diaphragm if the sky deflection approaches ~25% of the star-plus-sky
deflection.

Perhaps it should be stated explicitly that a star should always be *centered*
(and should remain centered) in the diaphragm when a deflection is made. This
is because, despite the fact that the purpose of the Fabry lens is to minimize
the variation produced as the star wanders around over the focal plane, there
always will be some change in photometric response as a function of how far off
the optical axis the star is. The best defense against this variation is to put all
stars (variable, comparison, check, standard, etc.) at the same place every time,
with the center of a diaphragm being the most easily reproducible location. An
illuminated crosshair, if your postviewer has one, can facilitate precise centering.
Thus, it is *not* a good idea to displace a star from the diaphragm center in an
attempt to exclude a nearby visual companion.

## 12.8   Recording the Time

Needless to say, you must know the time at which each photometric measure (or
sequence of measures) was made. This is because the light curve of any variable
is a plot of brightness versus *time*.

Generally, it is best to record either Universal Time (U.T.) or Julian date
(J.D.), realizing, of course, that the Julian date can be computed later from the

U.T. and the calendar date. You can determine the U.T. from an ordinary clock which displays your particular zone time (Z.T.), the relation between the two being simply

$$\text{U.T.} = \text{Z.T.} + \ell_z, \qquad (12.8.1)$$

where $\ell_z$ is the longitude of the time zone. For example, $\ell_z = 5$ hours for Eastern Standard or Central Daylight, $\ell_z = 6$ hours for Central Standard or Mountain Daylight, etc. If you have an ordinary clock already set later by $\ell_z$ hours, then U.T. can be read directly from that clock.

Generally, the time at which the variable is observed, not the comparison, is most important. If you have a strip chart recorder moving at a uniform rate, like an inch per minute, then the time of any deflection of any star can be determined later provided only that a few times are recorded on the chart paper, for example, every hour on the hour. Observers who turn off their chart recorder drive from time to time (to conserve paper) deprive themselves of this option.

For applying the differential atmospheric extinction correction you must know the hour angle at which the variable was observed. Hour angle can be read directly from an hour angle circle, if your telescope has one, or determined from the local sidereal time (L.S.T.) with the formula

$$h = \text{L.S.T.} - \alpha, \qquad (12.8.2)$$

where $h$ is hour angle and $\alpha$ is right ascension. Detailed instructions for computing your local sidereal time have been published in many places, for example, in the *Astronomical Almanac* each year and in the chapter by Bateson in Wood (1963), so we will not repeat them in this book. Reading the hour angle or recording the local sidereal time is not absolutely necessary, however, because it can be computed *ex post facto* from the U.T. or the Julian date. Boyd (1984) gives a simple one-line algorithm for doing this.

## 12.9 What to Write Down During Observing

In addition to recording the time, as discussed in the previous section, you need to record other information also, which either will be needed later to reduce the data or might be needed to investigate some unexpected problem which surfaces later.

### 12.9.1 The date

This simple item causes an unbelievable amount of confusion. Any observing night always includes the late evening hours of one calendar date followed by the early morning hours of the next calendar date. Some observers record the first, some the second, some whichever is applicable to the particular observation in question, which can be either. Some record the calendar date which goes along with the U.T. which they are recording, which changes when the U.T. goes from 23:59 to 00:01. Some record only the Julian date or its integral part, but this generates its own confusion because of the convention which has each new Julian day begin at noon in Greenwich rather than at midnight. Further complication has been introduced recently by the newly defined Modified Julian Date (M.J.D.),

$$\text{M.J.D.} = \text{J.D.} - 2,400,000.5, \qquad (12.9.1)$$

which *does* begin at midnight in Greenwich. We might mention here that nouveau era astronomique (N.E.A.) was introduced by Banachiewicz in 1922 and is used in some ephemerides of variable stars. With the starting date chosen to be at $0^h$ U.T. on January 1, 1801, you can use the relation

$$\text{N.E.A.} = \text{J.D.} - 2,378,860.5 \qquad (12.9.2)$$

to convert to Julian date.

Our recommendation is simple: at the beginning and/or the end of each observing night, record the *double calendar* date along with the year, for example, July 4–5, 1980. This is the best and simplest *completely unambiguous* way of identifying an observing night.

While not retracting the above recommendation, we point out that there are elegant one-line algorithms for computing the Julian date given the year, month, day, and time. Examples are Fernie (1983a) and Boyd (1984).

### 12.9.2   The photomultiplier or other photosensitive device

Many observers, in the course of their evolution, use a variety of different photodetectors. The particular one used on each and every night should be identified in some way meaningful to the observer. Examples would be 931A No. 2, 1P21 No. 1, EMI 9356 No. 1, 1P28 No. 3, Optec Unit B.

### 12.9.3   The voltage supplied to the photomultiplier

This is significant because of fatigue and saturation effects which can affect photomultipliers if a very bright star is observed at too high a voltage with a telescope of large enough aperture. The voltage can be changed during a night, depending on the star being observed, but all changes should be marked clearly.

### 12.9.4   The amplifier

The particular amplifier, pulse-counting device, or whatever, used on a given night should be identified unambiguously in some way. This is because the gain step calibrations (in the case of an operational amplifier) or the constants specifying the coincident-pulse corrections (in the case of pulse-counting equipment) are characteristic of the specific piece of apparatus being used and none other.

### 12.9.5   The gain steps

The gain steps used for each deflection of each star (or sky) with each filter must be specified unambiguously. This means specifying the coarse and the fine gain settings. If your gain steps are in magnitude units, then you should record the coarse and fine steps separately, not their sum. For example, record $5^m0 + 0^m5$, not simply $5^{m}5$. This is because $5^{m}5$ could have been formed with a coarse setting of $2^{m}5$ and a fine setting of $3^{m}0$ on the Stokes and Oliver amplifiers, and the calibration of the $2^{m}5$ coarse gain setting will differ from the $5^{m}0$ coarse gain setting.

### 12.9.6 The filter

For each individual deflection you can use abbreviated notations such as "v" and "b," but somewhere else it should be specified precisely which visual and blue filter was used. The catalogue description (such as 2mm of Schott GG–13 +1mm of Schott BG–12) might not be enough since individual filters of the same catalogue description can be different.

### 12.9.7 The diaphragm

Recording this can prove useful later on in deciding whether a certain nearby visual companion was included or excluded during photometry.

### 12.9.8 The star name

For each individual deflection you can use abbreviated notations such as "var," "comp," and "ck," but somewhere else the identification of each star should be specified unambiguously, either by one or two catalogue numbers or by an arrow marked on your finding chart.

### 12.9.9 Other remarks

You should make generous use of comments written directly on the chart recorder paper or other recording device output. Examples would be:

1. Cloud bank approaching from the northwest.

2. Quarter moon rising in the east.

3. Full moon in Scorpius.

4. Sky appears uniformly hazy.

5. Lightning on the western horizon.

6. Jet contrails crisscrossing the sky but not near the variable.

7. Star verified still near diaphragm center.

8. Star probably near edge of diaphragm.

9. Nearby star, barely visible, seen about 40 arcseconds north of the comparison star.

10. Intermittent patches of fog passing rapidly overhead.

11. U.T. clock $1^{m}07^{s}$ slow according to WWV.

12. Wrong star? Confusing finding chart.

**13.** Strong wind from northwest. Gusts up to 40 mph.

**14.** Seeing deteriorating. Images around 15 arcseconds in diameter.

**15.** Smoke from forest fire to the south.

**16.** Perfect sky. Best of the year.

### 12.9.10    Observing log

In addition to remarks made nightly you should maintain a log or journal in which you record relevant changes in the equipment such as dates on which the mirror was washed or realuminized, the photocell was removed from its socket and replaced, the amplifier was opened up and a resistor replaced, the filters were cleaned, etc.

## 12.10    Determining $k'$ on a Given Night

Even though differential photometry is such that the effect of differential atmospheric extinction is typically so small that $k'$ can usually be approximated crudely, sometimes just guessed, and occasionally ignored altogether, most photoelectric observers would like to know how to determine $k'$ at their observatory on a given night.

The time-honored way to determine the extinction coefficient is simply to observe a star at a variety of different zenith distances (as it rises towards the observer's celestial meridian or as it passes beyond his meridian or both). This star, called the extinction star, often is simply the comparison star used with the variable being observed on that night. Chapter 13 explains how the series of raw instrumental magnitudes of the extinction star are used to derive $k$ and $k'$.

Hardie (1959, 1962) has devised a very clever technique for determining extinction coefficients accurately and quickly. Speed is a useful advantage on nights which may be photometric at the beginning but then suddenly deteriorate later on without warning. In the space of a few tens of minutes you observe a few (at least two) standard stars, ranging from low in the sky (large X) to high in the sky ($X \approx 1$). Chapter 13 also explains how such measures of standard stars are used to derive $k$ and $k'$.

When selecting standard stars to use with the Hardie technique, you can choose from the lists of UBV standards described in Chapter 10 or from the "UBV Equatorial Extinction Star Network" of Crawford et al. (1971). This list of Crawford et al. includes 36 stars chosen carefully to be useful in a number of respects. They are all moderately bright, between V = $5^{m}4$ and V = $9^{m}1$, thus not too faint for most small telescopes and not too bright for most large telescopes. They are all very near the celestial equator, $-7° < \delta < +7°$ , and thus available to observers in both the Northern Hemisphere and the Southern. They are distributed over all 24 hours of right ascension so that some will be available during any season of the year. They are actually 18 red-blue pairs, most less than one degree apart in the sky and most well over one full magnitude different in B-V, so that the effect of $k'$ can be minimized while the value of $k''$ is being determined, if you choose not to assume a value.

It is usually adequate in differential photometry simply to guess or assume a value for $k'$ in each bandpass. It is possible to make surprisingly realistic guesses on individual nights simply by examining the sky visually. To do this well, you should have some feeling for the range of transparency encountered throughout the year at your own observatory and what value of $k'$ corresponds to the two extremes. This can be done by actually determining $k'$ on nights of such extremes. At Dyer Observatory, during a three-year interval on 30 nights during which the extinction was determined, Hardie (1959) found that $k_v$ ranged from a minimum of $0^m\!19$ on one night in the autumn to a maximum of $0^m\!64$ on one night in the summer. That unusually large value cannot be ignored, however, because it was determined on a night when the transparency was sufficiently uniform for the sky to have been deemed photometric. With extremes as a starting point, you could begin a simple chart relating to $k_v$ with visual impression of sky transparency conditions. Relevant factors might be prominence of the milky way, visibility or invisibility of certain stars known to be near 6th magnitude, altitude above the horizon where bright stars cease being visible, presence or absence of haze in the air as seen in a flashlight beam, etc. Needless to say, moonlight can complicate such estimates, but it is possible to make allowance for that.

Another approach is to use seasonal mean extinction coefficients. This will be meaningful at observatories which have certain seasons or times of the year when skies tend to be hazy, or dusty, or unusually clear, etc. At the Prairie Observatory in Illinois, Olson (1984) determined the extinction coefficient in the V bandpass numerous times over a 10–year interval. As his Figure 12–1 shows, one could assume an average value of $k_v$ depending on the month of observation and be confident that it would not be in error by more than $\pm 0.1$ magnitude/airmass in the winter or $\pm 0.2$ magnitude/airmass in the summer.

Armed with an estimate of $k'_v$ there are tricks for estimating the extinction coefficient in other bandpasses. One simple and not too unrealistic approach is to assume that any excess extinction over the minimum extinction for a normal clear sky, such as the one considered by Allen (1973, §55), is predominantly a result of haze, which affects all wavelengths pretty much equally. Thus, for example, if you measure or estimate $k_v = 0^m\!35$ on a certain night, whereas Allen (1973, §55) gives $k_v = 0^m\!20$, it is likely that $k'$ at all other wavelengths is greater by about the same amount, in this case $0^m\!15$. A refinement of this same approach is explained by Hardie (1959).

## 12.11 Determining Your Transformation Coefficients

Your transformation coefficients are best determined on a night of exceptionally uniform and constant transparency by observing a dozen or so standards of the photometric system (for example, the UBV system) which cover a broad range of color index (for example, $-0^m\!2 < \text{B-V} < +1^m\!2$) but which are relatively close together in the sky (for example, within a circle about 30° in diameter). They should be observed as swiftly as possible (for example, within an hour or so) and as near your zenith as possible (for example, within one hour before and one hour after crossing your celestial meridian).

Some one standard star, usually the one nearest the middle of the bunch, is selected as the reference star. This star will be used much as the comparison star is used in differential photometry of a variable star. Calling the reference star R and numbering the other standard stars 1 through 12, the sequence of deflections should be

$$R-1-2-3-R-4-5-6-R-7-8-9-R-10-11-12-R.$$

If time permits, you should immediately repeat the sequence, so that each standard star is observed twice. Needless to say, a sky deflection should immediately follow each star deflection, as was explained in Section 12.3 of this chapter. Chapter 13 will explain how transformation coefficients are derived from such data.

Just prior to beginning the sequence and, better still, also just after finishing the sequence, you should determine the extinction coefficients in each bandpass. The Hardie technique is best in this case because it is the fastest.

In selecting standard stars to observe for the determination of your transformation coefficients, an obvious choice, as pointed out by Hardie (1962), would be standard stars in a relatively bright open cluster. Stars in several clusters were observed at the same time the UBV system was set up originally and with the same equipment; consequently, those stars can be considered standards. Four such clusters which contain both blue (upper main-sequence) and red (red giant) stars are the Pleiades (Johnson and Morgan 1951), the Hyades (Johnson and Knuckles 1955), Praesepe (Johnson 1952), and IC 4665 (Johnson 1954).

A shortcut which is very attractive in many respects is to observe a star pair selected with transformation in mind. One example would be 27 and 28 LMi. They are only 15 arcminutes apart in the sky, sufficiently close that differential atmospheric extinction can be *ignored altogether* if they are observed within something like 20° of the zenith. They are sufficiently similar in brightness that you can observe them on exactly the same gain setting in the V and at least on the same coarse gain setting in the B and U. They are bright enough for small telescopes but not too bright for most large telescopes. And yet, they differ in color index by more than a full magnitude. Moreover, to our knowledge there is no evidence, published or unpublished, that either is variable. There are only two disadvantages in principle to using such a pair for determining transformation coefficients. First, although the differential magnitudes between 27 and 28 LMi on the UBV system have been determined, they are not bona fide standards of the UBV system. Second, by using only two stars, you are assuming implicitly that the transformation relation will be linear, which might not be so if your match to the UBV system is rather poor. In any case, the observing procedure is simply to intercompare the two on some one photometric night when they are as near the zenith as possible. In other words, the sequence of deflections would be

$$27-28-27-28-27-28-27-28-27-28-27-28-27,$$

the same as with normal differential photometry, imagining 28 LMi to be the variable and 27 LMi its comparison star. Such a sequence should yield six differential magnitude values, the average of which should be reliable to something like $\pm$ 0$^{\mathrm{m}}$005 or even better. Chapter 13 will explain how transformation coefficients are derived from such data. Hall (1983b, c) suggests four red-blue star pairs (one of which is 27 and 28 LMi) for use with this short-cut technique; they are all conveniently bright and there is one pair in each season of the sky.

The red-blue star pairs of Crawford et al. (1971) could be used in much the same way except that they will never rise quite to the zenith at observatories north or south of the equator. Moreover, as is the case with 27 and 28 LMi and as was pointed out by Crawford et al. (1971), they are not bona fide standards of the UBV system and they are just star *pairs*.

## 12.12  Determining Your Dead Time $\delta$ [1]

As explained in Chapter 5, you need to determine the dead time $\delta$ only if you use a pulse-counting photometer system. In general, the determination need be made only once for any given amplifier/discriminator and pulse counter. Although a value of $\delta$ is sometimes provided by the manufacturer of the electronics, our recommendation is to determine the value yourself. There are two basic strategies for doing so.

One involves the observation of stars of known magnitude in some standard photometric system, i.e., standard stars, but there are two ways of doing this. (1) You can determine $\delta$ simultaneously with the transformation coefficients and atmospheric extinction coefficients. In fact, this is the procedure recommended in the manual for the Cerro Tololo Inter-American Observatory. However, using the same observations to determine all three sets of coefficients gives greater uncertainty than is desirable. (2) Alternately, therefore, you can devise a procedure to isolate the dead time. An example would be to measure differentially a group of stars having a spread of $\sim$ 3 magnitudes, selecting these stars to be similar in color and close together in the sky so as to minimize the effects of differential transformation and differential atmospheric extinction, respectively. The very brightest of these stars must give enough counts for the dead time correction to be significant. This procedure often proves difficult in practice, mainly because it is difficult to select the optimum sample of stars.

Fortunately, with small telescopes, it is possible to use yet another technique which involves only two or three stars not necessarily standards of any photometric system. Those two or three stars should differ in brightness by $\sim$ 3 magnitudes and the brightest of them must produce a sizeable dead-time correction, but it is of secondary importance that they be very similar in color and very close together in the sky. The procedure is to measure the stars differentially with the full telescope aperture and then again very soon thereafter with an aperture stop which attenuates the starlight by a factor of $\sim$ 20×. A simple piece of cardboard, pierced with holes to let in $\sim \frac{1}{20}$ of the starlight and taped to the top of the telescope tube, makes an excellent stop. Various minor errors in this technique can be minimized if you take a few additional precautions.

1. Select stars of similar color.

2. Have a relatively large number of small holes distributed uniformly over the aperture, rather than a single larger hole.

---

[1] Written by Joel A. Eaton

**3.** Select stars close together in the sky and/or observe near the meridian and/or make your differential measures without the stop, with the stop, and again without the stop.

**4.** Apply the procedure to several different pairs of stars and on several different nights, to give a more accurate value of $\delta$ and to convince yourself that the value you adopt is likely valid.

The procedure just explained is somewhat similar to that described by Africano and Quigley (1977), except that theirs used an aperture stop *and* a neutral density filter.

In Chapter 13 we show how to use the observations thus obtained to determine your dead time $\delta$ and how to compute the true count rate $N$ from the observed count rate $n$.

# Chapter 13

# DATA REDUCTION

## 13.1 Determining Raw Instrumental Magnitude from the Deflections

The first step, very simply, is to measure the length of each deflection, i.e., each star + sky deflection and its corresponding sky deflection. If you are using a meter, each deflection has already been written down as a certain number of milliamps. If you are using digital recording equipment, each deflection is a certain number of counts. In the case of pulse counting (but not digital volt-meters) we need the true count rate $N$, derived from the observed count rate $n$ as explained in Section 13.12 of this chapter. If you are using a chart recorder, you can simply measure the length of each deflection with a ruler (in centimeters or millimeters, for example) or with whatever graduations are printed on the chart paper. To handle grass simply use your eye and brain as an analog computer to determine where a horizontal line should be drawn (actually or mentally) so that an equal amount of grass is above and below. Then measure the deflection at that level.

The next step, which is also very simple, is to determine the net star deflection by simple subtraction:

$$d(\text{star}) = d(\text{star} + \text{sky}) - d(\text{sky}), \tag{13.1.1}$$

where $d$ is deflection length. Notice that it is not necessary to ask at any point where the dark current level or the zero level was.

The next step is to adjust the net star deflection for whatever amplification was applied to it (if you are using an amplifier), and then convert to raw instrumental magnitude, $m$. If your amplifier has gain settings in factors ($1\times$, $2\times$, $3\times$, $5\times$, $10\times$, etc.), the relation is

$$\textit{Raw inst. Mag} = \qquad m = -2^{\text{m}}5 \log_{10} \frac{d}{f}, \tag{13.1.2}$$

where $f$ is the amplification factor which was applied to d(star). Of course, if the $3\times$ gain setting was found, by accurate calibration of the amplifier, to be amplifying by the factor 3.012, for example, then $f = 3.012$. If, instead, your amplifier has gain settings in magnitude units ($0^{\text{m}}0, 0^{\text{m}}5, 1^{\text{m}}0, 1^{\text{m}}5$, etc. for the fine settings and $0^{\text{m}}0, 2^{\text{m}}5, 5^{\text{m}}0$, etc. for the coarse settings), the relation is

$$m = -2^{\text{m}}5 \log_{10} d + g, \tag{13.1.3}$$

where $g$, in magnitude units, is the sum of the coarse and fine gain settings. If the coarse setting was $5^{\text{m}}0$, which amplifier calibration showed to be $5^{\text{m}}007$, for

189

example, and if the fine gain setting was 1ᵐ5, which can be assumed exact to
within ± 0ᵐ001, then $g = 6^m507$.

## 13.2   Determining Raw Instrumental Differential Magnitude

The first step is to take the average of or interpolate between the two comparison
star magnitudes on either side of the variable star magnitude. This is the magnitude the comparison star would have had if it hypothetically had been observed
simultaneously in time when the variable was. Then, by simple subtraction you
get

$Raw$ $Inst$ $Mag$ $diff =$ 
$$\Delta m = m(\text{var.}) - m(\text{interpol. comp.}), \tag{13.2.1}$$

where $\Delta m$ is the raw instrumental differential magnitude.

If the spacing in time between the various star deflections was not more or
less uniform, or if you were using an observing sequence in which more than one
variable star deflection was taken between two comparison star deflections, then
you have to use interpolation with respect to time in order to get the hypothetical
comparison star magnitudes needed. Although this is not really very difficult to
do on a pocket calculator, some observers prefer to do it graphically, either by
plotting the $m(\text{comp.})$ values versus time on graph paper and connecting them
with straight-line segments or by drawing straight-line segments, directly on the
chart recorder paper.

## 13.3   Determining Extraterrestrial Instrumental Differential
Magnitude

To correct for differential atmospheric extinction, in other words, to account for
the fact that light from the variable and its comparison have taken slightly different
paths through the atmosphere and thus suffered slightly different amounts of
extinction, you simply use equations like

$$\Delta v_o = \Delta v - k_v' \Delta X - k_v'' \bar{X} \Delta (B - V), \tag{13.3.1}$$

$$\Delta b_o = \Delta b - k_b' \Delta X - k_b'' \bar{X} \Delta (B - V), \tag{13.3.2}$$

or

$$\Delta u_o = \Delta u - k_u' \Delta X - k_u'' \bar{X} \Delta (U - B), \tag{13.3.3}$$

depending on which bandpass you are working with. Use the value of $k'$ measured
or assumed to be appropriate for that night and use the most appropriate value
of $k''$, both of which were discussed in Chapter 9.

$\Delta X$ and $\bar{X}$ are shorthand for

$$\Delta X = X(\text{var.}) - X(\text{interpol. comp.}) \tag{13.3.4}$$

and

$$\bar{X} = \frac{1}{2}[X(\text{var.}) + X(\text{interpol. comp.})]. \tag{13.3.5}$$

In differential photometry not too close to the horizon, it is entirely adequate
to use the secant approximation, Eq. (9.1.1), for $X$. Because $m(\text{var.})$  and

$m$(interpol.comp.) are assumed to have been taken at the same time, it is convenient to use the fact that

$$h(\text{interpol. comp.}) = h(\text{var.}) + \alpha(\text{var.}) - \alpha(\text{comp.}), \qquad (13.3.6)$$

which follows from Eq. (12.8.2) since the sidereal times for $m$(var.) and $m$(interpol.comp.) are the same.

It is somewhat embarrassing that you need $\Delta$(B–V) at this stage but do not know it yet. It is not really a serious problem, though. If you are doing two-filter photometry, then $\Delta$(b–v) can be used instead because it rarely differs enough from $\Delta$(B–V) to make a significant change in the $k''$ term. If you are doing single filter photometry, there are two choices. If the variable has been observed before by others, with the same comparison star, then the value of $\Delta$(B–V) at various phases of its light curve is probably known already. Or, if worse comes to worst, you can always determine $\Delta$(B–V) yourself later.

## 13.4 Determining the Standard Differential Magnitude

To transform your extraterrestrial instrumental differential magnitudes to differential magnitudes on the standard system you are trying to match, you simply use equations like

$$\Delta V = \Delta v_o + \epsilon_v \Delta(B - V), \qquad (13.4.1)$$

$$\Delta B = \Delta b_o + \epsilon_b \Delta(B - V), \qquad (13.4.2)$$

or

$$\Delta U = \Delta u_o + \epsilon_u \Delta(U - B), \qquad (13.4.3)$$

depending on which bandpass you are working with. Use the values for the transformation coefficients $\epsilon_v$, $\epsilon_b$, or $\epsilon_u$ which you have determined as appropriate for your instrumental system, as was explained in Chapter 10.

Again, you have the slight embarrassment of needing $\Delta$(B–V) and $\Delta$(U-B) before you know them but, similar to what was done before, you can simply use $\Delta(b-v)_o$ and $\Delta(u-b)_o$ instead, since they rarely differ enough from $\Delta$(B–V) and $\Delta$(U–B) to make a significant change in the last term.

## 13.5 The Heliocentric Correction

Any time written on the chart paper alongside a star deflection (whether it be a local time, U.T., Julian Date, sidereal time, or whatever) is implicitly a geocentric time, the time when the light signal in question reached the earth. It is customary in variable star photometry to add the heliocentric correction to each geocentric time to give the corresponding heliocentric time, i.e., the time when that same light signal reached the sun.

A little thought tells us why this is done. As Copernicus told us over 400 years ago, it is the earth which goes around the sun, not vice versa. Depending on where the earth is in its orbit and where the particular star is located, the wave front of light from that star can reach the earth up to $8^m19^s$ before it reaches the sun or up to $8^m19^s$ later. As you probably remember, $8^m19^s$ is the time is takes light, traveling 186,000 miles/second, to traverse the 93,000,000-mile distance between the earth and the sun. Thus, for any given star, the heliocentric correction can be negative, positive, or zero depending on the day of the year. The maximum heliocentric correction (positive or negative) for any given star, however, will be less than $8^m19^s$ except for stars which happen to lie exactly in the ecliptic plane, i.e., the plane of the earth-sun orbit. Similarly, any star which happens to be exactly in the direction of the ecliptic north or south pole will have a heliocentric correction which is exactly zero through the year. The equation is

$$\text{hel. corr.} = -KR[\cos\theta\cos\alpha\cos\delta + \sin\theta(\sin\epsilon\sin\delta + \cos\epsilon\cos\delta\sin\alpha)], \quad (13.5.1)$$

where

$K$ = the time for light to travel one astronomical unit = $8^m19^s$ = $0^d0057755$,

$R$ = the actual distance between earth and sun in astronomical units,

$\theta$ = the longitude of the sun,

$\epsilon$ = the obliquity of the ecliptic = $23°27'$,

$\alpha$ = the star's right ascension, and

$\delta$ = the star's declination.

Note that $R$ and $\theta$ depend on the date of observation, $\alpha$ and $\delta$ depend on the star being observed, and $K$ and $\epsilon$ are fixed. One of three methods is usually used to determine the heliocentric correction.

One method involves looking up $R$ and $\theta$ in the *Astronomical Almanac* for the observing night in question and entering them into Eq. (13.5.1). This method is rather tiresome unless the equation is made part of a computer program.

A second method is to use the convenient tables of heliocentric corrections prepared by Landolt and Blondeau (1972). These tables give the heliocentric correction in units of $0^d0001$ as a function of date in the year (in 10-day intervals), declination of the star (in 10° intervals), and right ascension of the star (in $1^h$ intervals). The only minor inconvenience is having to do a three-way interpolation.

Many people, anxious to perform data reduction automatically as much as is possible with a microcomputer, find ways to apply Eq. (13.5.1) in the form of a simple but adequate algorithm. One such algorithm for the heliocentric correction is given by Boyd (1984).

It is customary to apply the heliocentric correction to Julian date, rather than to U.T. and calendar date. Thus, we have

$$\text{JD (hel.)} = \text{JD (geo.)} + \text{hel. corr.} \quad (13.5.2)$$

As the equation indicates, the heliocentric correction (which itself can be either a positive or a negative quantity) is *added* to the geocentric time. Because most stars are observed at night, when the star and the sun are on opposite sides of the earth, it follows logically that most computed heliocentric corrections will be positive.

## 13.6   How to Compute Phase

If a star varies in light periodically rather than irregularly or randomly, then you can think of each cycle as a more or less identical repeat performance. Progress through each cycle is described by phase, a number which is directly proportional to time and goes from 0 to 1 or from 0° to 360° . Customarily, degrees are used only when dealing with a variable, like an eclipsing binary, which involves orbital motion; in such a case the line of centers connecting the two stars actually does sweep out an angle, which covers 360° during each orbital revolution.

For a periodic variable (or any periodic phenomenon) you can use an ephemeris to determine times when the variable is at a certain point of its cycle. Heliocentric Julian date is almost always used for this purpose. Let us take the eclipsing binary BM Orionis as a specific example and write the ephemeris for times of minimum light as

$$\text{JD (hel.) minimum light} = 2,440,265.343 + 6.470525\,n. \qquad (13.6.1)$$

Here, the first number is the initial epoch, often designated $t_o$, some particular time in the past when BM Orionis did drop to minimum light. The second number is the period, usually designated $P$, the amount of time between successive occurrences of minimum light. In this case of an eclipsing binary, the period is also the amount of time required for one complete orbital revolution. The number n is the cycle number (sometimes called E, the epoch number), an integer which can be negative to identify times of minimum light which occurred before the initial epoch or positive to identify times of minimum light which occurred after the initial epoch. Thus, very simply, to make a list or ephemeris of all times when BM Orionis did reach or will reach minimum light, simply insert all possible integers, both negative and positive, and perform the arithmetic.

For an eclipsing binary, the initial epoch is customarily taken to be a time of minimum light, i.e., the middle of the primary (deeper) eclipse. For a Cepheid or an RR Lyrae variable, the initial epoch is usually taken to be a time of maximum light. For a spectroscopic binary with an eccentric orbit, it is usually taken to be a time of periastron passage, i.e., a time when the two stars are closest together. For a spectroscopic binary with a circular orbit there is no periastron so something else must be used, like a time of maximum radial velocity, a time of conjunction, or a time of ascending node.

For a certain Julian date at which you make a particular observation of a variable, you compute the phase of that observation very simply with the formula

$$\text{Phase} = \frac{\text{JD(hel.)} - t_o}{P}. \qquad (13.6.2)$$

Let us use BM Orionis again as an example. The phase of an observation made on JD(hel.) 2,444,123.789 would be

$$\text{Phase} = \frac{2,444,123.789 - 2,440,265.343}{6.470525} = 596.311. \qquad (13.6.3)$$

This tells you that your observation was made at a time when BM Orionis had completed 596 revolutions or cycles since the initial epoch and was 31.1% of the way through the 597th cycle. In plotting observations on a light curve (light vs. phase) which combines observations from several different cycles, you ignore the integral part, in this case the 596, and use only the fractional part, in this case the 0.311.

## 13.7   How to Summarize Your Data

If you are publishing your own data or if you are transmitting your data to someone else in fully reduced form, the essential information for each individual observation usually is just heliocentric Julian date and differential magnitude corrected for differential atmospheric extinction and differential transformation. If the variable is periodic and an ephemeris is available, you can compute phase for each observation, but this is a convenience rather than a necessity. An example is shown in Table 13–4. It is better to leave the magnitudes differential rather than adding on the magnitude of the comparison star, even if it is known, but do not forget to specify which comparison star you used.

If you are transmitting your data to the coordinator of a cooperative observing campaign, he may prefer to perform many of the corrections himself, for the sake of consistency and to minimize the likelihood of computational errors. Also, by studying the raw magnitudes in detail, he can sometimes notice places where something was going wrong. In Table 13–5, we suggest a format for reporting data, one which lets the observer reduce his data up to the point of raw instrumental differential magnitudes but lets the coordinator apply the heliocentric correction, compute phases, correct for atmospheric extinction, and transform to the standard photometric systems. To do his part, the coordinator needs to know only the equatorial coordinates for the variable and its comparison star, the observer's latitude and longitude, an estimate of the extinction coefficients appropriate for each night, and the observer's transformation coefficients.

## 13.8   A Numerical Example for Reducing Differential Photometry

We have included a detailed, reasonably realistic, self-consistent numerical example to illustrate the procedures explained in the first seven sections of this chapter, from deflections to final summary sheet. Table 13–1 accompanies Sections 13.1 and 13.2, Table 13–2 goes with Sections 13.3 and 13.4, Table 13–3 with Section 13.5, and Table 13–4 with Sections 13.6 and 13.7. In this numerical example we have made the following assumptions:

1. The observatory site is at $\phi = +30°$ and $\ell = 5^h50^m$ west.

2. The check star is at $\alpha\ 4 = 12^h12^m$, $\delta = +31°00'$; the comparison star is at $\alpha = 12^h12^m$, $\delta = +30°00''$; and the variable is at $\alpha = 12^h12^m$, $\delta = +29°00'$.

3. The observing night was July 4–5, 1980.

4. A diaphragm 30 arcseconds in diameter was used until U.T. = 2:36, after which time a diaphragm 60 arcseconds was used, because of deteriorating seeing conditions.

5. Deflections on all three stars with both filters used the same coarse gain setting, 10 megohms, but different fine gain settings: 1×, 3×, and 5×. Calibration of the fine gain settings showed them to be amplifying by factors of 1.000, 3.007, and 4.991, respectively.

6. After the amplifier was balanced, the chart recorder was offset by about 3 mm, to avoid off-scale negative readings in case the amplifier lost its balance.

**7.** The chart recorder has a full-scale range of 10 inches, a bit more than 25 cm.

**8.** Values assumed for the extinction coefficients were $k'_v = 0^m3$, $k''_v = 0^m00$, $k'_b = 0^m5$, and $k''_b = -0^m03$.

**9.** The transformation coefficients used were $\epsilon_v = +0.04$ and $\epsilon_b = -0.01$.

**10.** Phases for the variable were computed with the ephemeris JD(hel.) 2,440,123.4567 + $1^d234567$ n.

*(handwritten: $= Comp$ ; Avg of Comp ; $-2.5 \log d/f$ ; Raw Inst Mag diff)*

### TABLE 13–1
### REDUCTION TO
### RAW INSTRUMENTAL DIFFERENTIAL MAGNITUDE

| U.T. | Star | Filter | d(star +sky) | d (sky) | d (star) | gain | f/Gain | m (interpol) | m | Δm |
|------|------|--------|------|------|------|------|------|------|------|------|
| 2:03 | comp | v | 0.204 | 0.005 | 0.199 | 10M,1× | 1.000 | 1.753 | | |
| 2:05 | comp | b | .188 | .006 | .182 | 1× | 1.000 | 1.850 | | |
| 2:08 | check | v | .183 | .013 | .170 | 5× | 4.991 | 3.669 | 1.758 | 1.911 |
| 2:10 | check | b | .170 | .012 | .158 | 3× | 3.007 | 3.199 | 1.859 | 1.340 |
| 2:13 | comp | v | .202 | .005 | .197 | 1× | 1.000 | 1.764 | | |
| 2:15 | comp | b | .185 | .006 | .179 | 1× | 1.000 | 1.868 | | |
| 2:18 | var | v | .208 | .013 | .195 | 5× | 4.991 | 3.520 | 1.770 | 1.750 |
| 2:20 | var | b | .237 | .018 | .219 | 5× | 4.991 | 3.394 | 1.871 | 1.523 |
| 2:23 | comp | v | .200 | .005 | .195 | 1× | 1.000 | 1.775 | | |
| 2:25 | comp | b | .184 | .006 | .178 | 1× | 1.000 | 1.874 | | |
| 2:28 | var | v | .211 | .013 | .198 | 5× | 4.991 | 3.504 | 1.780 | 1.724 |
| 2:30 | var | b | .240 | .018 | .222 | 5× | 4.991 | 3.380 | 1.880 | 1.500 |
| 2:33 | comp | v | .198 | .005 | .193 | 1× | 1.000 | 1.786 | | |
| 2:35 | comp | b | .182 | .006 | .176 | 1× | 1.000 | 1.886 | | |
| 2:37 | comp | v | .204 | .011 | .193 | 1× | 1.000 | 1.786 | | |
| 2:39 | comp | b | .192 | .015 | .177 | 1× | 1.000 | 1.880 | | |
| 2:42 | var | v | .240 | .043 | .197 | 5× | 4.991 | 3.509 | 1.778 | 1.731 |
| 2:44 | var | b | .172 | .039 | .133 | 3× | 3.007 | 3.386 | 1.874 | 1.512 |
| 2:47 | comp | v | .207 | .011 | .196 | 1× | 1.000 | 1.769 | | |
| 2:49 | comp | b | 0.194 | 0.015 | 0.179 | 10M,1× | 1.000 | 1.868 | | |

*(handwritten: $(3.669 - 1.758)/7 = 1.911$ ; Comps ; $(3.520 - 1.770)/? = 1.75$ ; (B–V))*

### TABLE 13–2
### APPLYING DIFFERENTIAL CORRECTIONS
### FOR
### EXTINCTION AND TRANSFORMATION

| U.T. | Star | h | ΔX | $\overline{X}$ | Δv | $k'_v\Delta X$ | $k''_v\Delta(B–V)$ | $\Delta v_0$ | $\epsilon_v\Delta(B–V)$ | ΔV |
|------|------|------|------|------|------|------|------|------|------|------|
| 2:09 | chk | +3:00 | −0.0034 | 1.28 | 1.911 | −0.001 | 0.000 | 1.912 | −0.023 | 1.889 |
| 2:19 | var | +3:10 | +0.0045 | 1.32 | 1.750 | +0.001 | 0.000 | 1.749 | −0.009 | 1.740 |
| 2:29 | var | +3:20 | +0.0053 | 1.37 | 1.724 | +0.002 | 0.000 | 1.722 | −0.009 | 1.713 |
| 2:43 | var | +3:34 | +0.0066 | 1.44 | 1.731 | +0.002 | 0.000 | 1.729 | −0.009 | 1.720 |

| U.T. | Star | h | ΔX | $\overline{X}$ | Δb | $k'_b\Delta X$ | $k''_b\Delta(B–V)$ | $\Delta b_0$ | $\epsilon_b\Delta(B–V)$ | ΔB |
|------|------|------|------|------|------|------|------|------|------|------|
| 2:09 | chk | +3:00 | −0.0034 | 1.28 | 1.340 | −0.002 | +0.022 | 1.320 | +0.006 | 1.326 |
| 2:19 | var | +3:10 | +0.0045 | 1.32 | 1.523 | +0.002 | +0.009 | 1.512 | +0.002 | 1.514 |
| 2:29 | var | +3:20 | +0.0053 | 1.37 | 1.500 | +0.003 | +0.009 | 1.488 | +0.002 | 1.490 |
| 2:43 | var | +3:34 | +0.0066 | 1.44 | 1.512 | +0.003 | +0.010 | 1.499 | +0.002 | 1.501 |

| TABLE 13–3 COMPUTING HELIOCENTRIC JULIAN DATE | | | | |
|---|---|---|---|---|
| U.T. | Star | JDf (geo.) | Hel.corr. | JDf(hel.) |
| 2:09 | check | 0.5896 | −0.0020 | 0.5876 |
| 2:19 | var | 0.5965 | ″ | 0.5945 |
| 2:29 | var | 0.6035 | ″ | 0.6015 |
| 2:43 | var | 0.6132 | ″ | 0.6112 |

| TABLE 13–4 FINAL SUMMARY OF DATA | | | | |
|---|---|---|---|---|
| JD(hel.) | Star | Phase | $\Delta V$ | $\Delta B$ |
| 2,444,425.5876 | check | – | $1^m889$ | $1^m326$ |
| .5945 | var | $0^P7342$ | $1^m740$ | $1^m514$ |
| .6015 | var | $0^P7389$ | $1^m713$ | $1^m490$ |
| 2,444,425.6112 | var | $0^P7477$ | $1^m720$ | $1^m501$ |

You may wish to note the following points:

1. Comparison star deflections were repeated when the diaphragms were changed.

2. When the larger diaphragm was used, the sky deflections and the star + sky deflections both became larger, especially noticeable for the variable, which was amplified more than the comparison star.

3. The larger star + sky deflection produced by the diaphragm change made the blue-filter variable deflection exceed the 25–cm chart recorder range, so the gain was lowered from 5× to 3×.

4. In Table 13–1, deflection lengths are recorded in meters (rather than in mm or cm) simply to keep all raw magnitudes positive.

5. Although the visual- and blue-filter deflections were taken a couple of minutes apart, further reduction considers them made at the same time, an acceptable approximation for a relatively slow variable.

6. $h$, $\Delta X$, and X in Table 13–2 are computed at times when the variable (or the check star) was observed, not when the comparison star was observed.

7. The $k_b'$ term in Table 13–2 is quite small, compared to the $k_b''$ term, for example.

8. The $k'$ and $k''$ terms in Table 13–2 have been subtracted because they appear in Eqs. (11.2.1), (11.2.2), and (11.2.3) with negative algebraic signs. The $\epsilon$ terms have been added since they appear in Eqs. (11.2.7), (11.2.8), and (11.2.9) with positive algebraic signs.

## 13.9   Determining $k'$ on a Given Night

In Section 12.10 of Chapter 12, it was explained that differential photometry typically does not require actual determination of $k'$ on a given night. If, however, you do observe extinction stars for this purpose, the data are reduced as follows.

If you were using the single extinction star method, you simply plot each raw instrumental magnitude $m$ versus its airmass $X$. Such a plot is illustrated in Fig. 13–1 and the data upon which it is based are given in Table 13–6. If

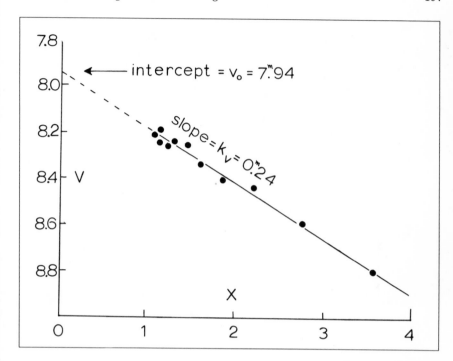

**Figure 13–1.** Raw instrumental magnitude versus airmass plot, for determination of extinction coefficient.

the sky transparency held constant while these observations were being obtained, then the points should form a straight line. The slope of this line, in units of magnitude/airmass, will be the extinction coefficient $k$. In the illustration $k = 0^m24$. The intercept of the line, the value of m where an extrapolation of the line reaches $X = 0$, is the extinction star's extraterrestrial instrumental magnitude $m_o$. In the illustration $v_o = 7^m94$. This can be seen by examining Eq. (9.1.5) and noting that it is the equation of a straight line.

If the line shows any curvature or if the points taken as the star was rising do not blend in with the points taken as the star was setting, that is an indication that the sky transparency must not have held constant, the night was not photometric, and there is essentially no way to use those data to derive any meaningful information about $k$. The last column in Table 13–6 shows how far each observed V value lies above or below the straight line given by the equation $v = v_o + 0^m24X$. Running your eye down that column you see no systematic trends in the residuals as a function of time, hence no indication of time-dependent transparency.

Of course, the slope of the line is $k$, not $k'$. To derive $k'$ by application of Eq. (9.1.6) you must know the color index B–V of the extinction star. Usually this is no problem because often the star is a standard of the UBV system or has its UBV magnitudes published in some catalogue. Or it will be the comparison

$$k_\lambda = k'_\lambda + k''_\lambda (B-V)$$

TABLE 13–5
SUGGESTED FORMAT FOR REPORTING DATA

| λ And, ψ And | | | | | July 6, 1976 | | |
| Lovell, Hickox | | | | | | | |
| 1976 | | 244+ | | –850 Volts | | | |

| Month Date | U.T. | JD(geo.) | Hel. corr. | Filter | $\Delta m$ | dia. | Notes |
|---|---|---|---|---|---|---|---|
| May 17–18 | 6:24 | 2916.767 | | V | –1.095 | 30" | |
| | 6:30 | .771 | | " | –1.101 | " | |
| | 6:35 | .774 | | " | –1.102 | " | |
| 24–25 | 7:02 | 2923.793 | | V | –1.123 | 60" | a |
| | 7:06 | .796 | | " | –1.146 | " | |
| | 7:09 | .798 | | " | –1.134 | " | |
| | 7:12 | .800 | | " | –1.128 | " | |
| | 7:15 | .802 | | " | –1.138 | " | |
| 29–30 | 6:58 | 2928.790 | | V | –1.101 | 30" | |
| | 7:03 | .794 | | " | –1.112 | " | |
| | 7:05 | .795 | | B | –1.598 | " | |
| | 7:09 | .798 | | " | –1.603 | " | |
| June 3–4 | 6:45 | 2933.781 | | V | –1.166 | " | |
| | 6:50 | .785 | | " | –1.169 | " | |
| | 6:55 | .788 | | " | –1.164 | " | b |
| 7–8 | 8:00 | 2937.833 | | V | –1.113 | " | c |
| | 8:05 | .837 | | " | –1.108 | " | |
| | 8:12 | .842 | | " | –1.110 | " | |
| | 8:15 | .844 | | " | –1.099 | " | |

a - Full moon but sky deflections not too big
b - Clouds seen in sky but deflections repeat OK
c - Almost in the trees but deflections seem reasonable

TABLE 13–6
DETERMINING THE
EXTINCTION COEFFICIENT

| U.T. | $h$ | $X$ | $v$ | $v_o + 0.24X$ | diffr. |
|---|---|---|---|---|---|
| 1:10 | –5:00 | 1.858 | 8ᵐ41 | 8ᵐ386 | + 0ᵐ024 |
| 2:30 | –3:40 | 1.440 | 8.26 | 8.386 | –0.026 |
| 3:50 | –2:20 | 1.225 | 8.26 | 8.286 | –0.026 |
| 5:10 | –1:00 | 1.124 | 8.18 | 8.210 | –0.030 |
| 6:30 | +0:20 | 1.106 | 8.20 | 8.205 | –0.005 |
| 7:50 | +1:40 | 1.163 | 8.25 | 8.219 | +0.031 |
| 9:10 | +3:00 | 1.314 | 8.24 | 8.255 | –0.015 |
| 10:30 | +4:20 | 1.614 | 8.34 | 8.327 | +0.013 |
| 11:50 | +5:40 | 2.208 | 8.44 | 8.470 | –0.030 |
| 12:30 | +6:20 | 2.730 | 8.58 | 8.595 | –0.015 |
| 13:10 | +7:00 | 3.559 | 8.80 | 8.794 | +0.006 |

star for your variable star program, in which case you probably have found or determined its B–V. If not known beforehand, you can simply determine its B–V *ex post facto* on some other night.

If you are using the Hardie technique, the reduction procedure is somewhat similar but a bit more complex. Taking V of the UBV system as an example and combining Eq. (11.2.1) with (11.2.7), we get

$$V = v - k'_v X - k''_v X(B - V) + \epsilon_v(B - V) + Z_v. \qquad (13.9.1)$$

This can be arranged to give

$$v - V - k''_v X(B - V) + \epsilon_v(B - V) = k'_v X - Z_v, \qquad (13.9.2)$$

which, in a plot of the quantity on the left versus $X$, is the equation of a straight line for which the slope is $k_v$ and the intercept at $X = 0$ is $Z_v$ with the opposite algebraic sign. Each standard star observed gives one point on the plot. Notice that $v$ is measured and $X$ is derived from the hour angle recorded each time one standard is observed; V and B–V are known because the stars are UBV standards; and $k''_v$ and $\epsilon_v$ are previously determined quantities which do not change from night to night at a given observatory.

## 13.10  Determining Your Transformation Coefficients

If you are using the method based on a dozen or so standard stars, the procedure is as follows. We take V of the UBV system as an example. Equation (13.9.1) can be written down twice, once for some one of the standards and again for the reference standard, and subtracted to give

$$\Delta V = \Delta v - k'_v \Delta X - k''_v \bar{X} \Delta(B - V) + \epsilon_v \Delta(B - V), \qquad (13.10.1)$$

where $\Delta$ is in the sense standard minus reference. This can be rearranged to give

$$\Delta V - \Delta v + k'_v \Delta X + k''_v \bar{X} \Delta(B - V) = \epsilon_v \Delta(B - V). \qquad (13.10.2)$$

If the quantity on the left is plotted versus $\Delta(B-V)$, the result should be a sequence of points falling in a straight line, the slope of which will be $\epsilon_v$. Notice that there is no intercept in this case; the line *must* pass through the $0, 0$ origin of the plot.

We must, however, explain exactly how each point is computed. First, for each measure of the reference star, plot on a separate graph all of the raw instrumental visual magnitudes $v$ of the reference star versus time, for example, versus U.T. Then connect those points with straight-line segments. Also, on another graph plot all of the airmass values $X$ of the reference star versus time. Then, by simple interpolation, determine the values of $v$ and $X$ at whatever times the other standard stars were observed. Thus, the quantities $\Delta v$, $\Delta X$, and $\bar{X}$ are given by

$$\Delta v = v(\text{std.}) - v(\text{interpol. ref.}), \qquad (13.10.3)$$

$$\Delta X = X(\text{std.}) - X(\text{interpol. ref.}), \qquad (13.10.4)$$

and

$$\bar{X} = \tfrac{1}{2}[X(\text{std.}) + X(\text{interpol. ref.})]. \qquad (13.10.5)$$

Since the reference star and all of the standards are UBV standards, $\Delta V$ and $\Delta(B-V)$ result from subtraction of the corresponding published UBV magnitudes. The appropriate value to use for $k'_v$ is the value which you determined on that night (or the average of the two values, if you determined it before and after you observed the standards). The value of $k_v''$ was determined or assumed previously for your photometric equipment and should not change from night to night.

If you are using the star pair method, the procedure is much simpler, because you have just one difference to work with. Again, we use V of the UBV system as an example. In this case, solving Eq. (13.10.2) for $\epsilon_v$, we get

$$\epsilon_v = \frac{\Delta V - \Delta v + k_v'\Delta X + k_v''\bar{X}\Delta(B-V)}{\Delta(B-V)}. \tag{13.10.6}$$

Here $\Delta v$ is the average raw instrumental differential magnitude based on the half-dozen or so inter-comparisons, $\Delta X$ is the average differential airmass between the two, and $\bar{X}$ is the average airmass of the two. Both $\Delta V$ and $\Delta(B-V)$ must, of course, be accurately known quantities for such a transformation pair. For 27 and 28 LMi, we have found $\Delta V = 0^m378$ and $\Delta(B-V) = -1^m03$, where $\Delta$ is in the sense 27 minus 28. And $k_v''$ is considered previously determined or assumed for your photometric equipment. If the star pair was observed within $\sim 20°$ of your zenith, then the $k'$ term will be so small as to be insignificant. The $k''$ term, on the other hand, will be sizable for bandpasses, like blue, which do have a measurable value of $k''$.

## 13.11   Determining Your Dead Time $\delta$

The basic idea behind the so-called dead-time correction is that, at high count rates, the recorded count rate $n$ underestimates the true rate $N$ because some pulses are too close to each other to be resolved by the pulse counter. Assume that, every time the amplifier/discriminator detects a pulse, it must wait the time $\delta$ before it can detect another pulse. If the rate at which pulses are being produced is $N$ counts per second, the number of pulses that will arrive during this dead time is $\delta N$. So, for each photon detected, $\delta N$ photons are missed and the rate at which photons are actually counted is

$$n = N(1 - \delta N). \tag{13.11.1}$$

This gives us three useful approximation formulas which may be used to correct for this effect:

$$N = \frac{n}{1 - \delta N} \approx \frac{n}{1 - \delta n} \approx n(1 + \delta n) \tag{13.11.2}$$

Fernie (1976) discusses higher-order approximations to the dead-time correction.

Let us assume you have used the aperture-stop technique with two stars: A and B. With no stop the count rates are $n_A$ and $n_B$ ; with the stop the count rates are $n_A'$ and $n_B'$. The corresponding magnitudes corrected for dead time would be

$$m_A = -2^m5 \log \frac{n_A}{1 - \delta n_A}, \tag{13.11.3}$$

$$m_B = -2^m5 \log \frac{n_B}{1 - \delta n_B}, \tag{13.11.4}$$

$$m_A' = -2^m5 \log \frac{n_A'}{1 - \delta n_A'}, \tag{13.11.5}$$

and

$$m'_B = -2\overset{m}{.}5 \log \frac{n'_B}{1 - \delta n'_B}. \tag{13.11.6}$$

Here we have used the middle of the three forms of Eq. (13.11.12). The correct value of $\delta$ would be that for which

$$m_A - m_B = m'_A - m'_B \tag{13.11.7}$$

and can be determined by a little numerical trial and error.

## 13.12  Applying The Dead-Time Correction

Notice that, in *determining* $\delta$, we have used the middle of the three forms of Eq. (13.11.2). Therefore, in *applying* the dead time correction, you must use that same form, i.e.,

$$N = \frac{n}{1 - \delta n}, \tag{13.12.1}$$

to convert your observed count rates $n$ into correct count rates $N$. We mention here that both $n$ and $N$ are customarily in units of counts per second, in which case the dead time $\delta$ must be in units of seconds.

# Chapter 14

# ACHIEVING MAXIMUM ACCURACY

## 14.1 Introduction

It could be argued that any photometry is better than no photometry, and that more photometry is better than less photometry, but all would agree that the ultimate value of photometry depends strongly on its accuracy. The purpose of this chapter is to identify the major factors which can contribute to inaccuracies in photoelectric photometry.

Many of these have been alluded to already but we wish here to make the following point. Your photometric equipment can be functioning, or at least not malfunctioning in any gross way, and you can be obtaining results which can be considered meaningful photoelectric photometry, yet a number of factors can be producing small errors which singly or collectively will render your photometry far less useful than it potentially could be. We think blue ribbon photoelectric photometrists, professional or amateur, should

1. Know what the potential sources of error are,

2. Have a good feeling for how those errors enter in and approximately how large each one can be,

3. Be able to recognize the symptoms of each one when and if it creeps in, and

4. Have the capability of eliminating or minimizing each of them one by one.

The ultimate goal is to reach a point where the accuracy of your photometry is limited by the factors over which you, the observer, have no control as long as you work with your particular telescope at your particular observing site: scintillation noise, photon noise, and slight variations in sky transparency.

By way of perspective, we will draw on two decades of experience with photoelectric photometry, our own and that of scores of other photoelectric photometrists we have collaborated with, which have shown what types of errors in practice most commonly do occur and do actually affect accuracy. There are other perspectives and much more has been said about the subject. The published proceedings of the "Workshop on Improvements to Photometry," edited by Borucki and Young (1984), would make an excellent supplemental reference.

For each source of error we will assume a typical or representative size for it and show how the final differential magnitude is affected. In general, an error twice as large or half as large will affect the differential magnitude twice as much or half as much. It is convenient to use the approximation that

$$2^m5 \log_{10}(1 + \epsilon) \approx \epsilon \quad \text{if} \quad \epsilon \ll 1. \tag{14.1.1}$$

For example, if the ratio in brightness between two light sources is 1.02, then the difference in their magnitude will be very nearly $0^m02$. Actually, this approximation is surprisingly good and hence useful up to almost a half magnitude. A pocket calculator shows that brightness ratios very near 1.0 give magnitude differences only 9% too large; for a brightness ratio of 1.18 the approximation is exact; and a brightness ratio of 1.4 gives a magnitude difference only 9% too small.

## 14.2   Random Errors, Systematic Errors, and Gross Errors

There are basically three types of errors in any experimental science: random errors (sometimes called accidental errors or Gaussian errors), systematic errors, and gross errors. The first two are often referred to as internal errors and external errors, respectively. Moreover, the distinction between precision and accuracy is to some extent the distinction between random and systematic errors. Some examples will illustrate how the three differ from each other.

Using the recommended basic sequence of deflections described in Chapter 12 would result in three separate estimates of $\Delta V$. Several effects (slightly variable atmospheric transparency, scintillation, photon noise, somewhat inexact measuring of the grassy deflections, etc.) will cause the three values of $\Delta V$ to differ slightly from each other. The standard deviation of the three from their mean, $\sigma_1$, would be a measure of the random error.

Other effects (imprecisely determined $\epsilon_v$, inappropriate value of $k'$ assumed, an old amplifier calibration no longer valid, etc.) would affect all three values almost exactly the same and hence would not manifest themselves in $\sigma_1$. These would be systematic errors. Systematic errors, because they do not reveal themselves directly, are more difficult to deal with but are present nevertheless and hence cannot be ignored. Systematic errors often do reveal themselves, however, when observations of the same star made at different observatories are combined on the same light curve. This is because the corresponding systematic errors at the other observatories probably will be different, either larger or smaller or in the opposite direction. But some systematic errors might never reveal themselves. For example, if the variable is a very bright star, much brighter than the comparison star, its strong light could be saturating the photomultiplier and thus its brightness be underestimated. If all observers had telescopes of comparable aperture and operated their 1P21 photomultiplier tubes at about the same voltage, then all would obtain $\Delta V$ values which were too faint by about the same amount and the error would not be readily apparent.

Gross errors are the result of some actual mistake and usually produce an extremely large error. Examples would be

1. Reading a meter or chart recorder deflection as 78.3 instead of 68.3,

2. Misidentifying the comparison star or the variable and observing the wrong star,

3. An arithmetic error somewhere in the data reduction process,

4. Forgetting to subtract the sky deflection,

5. Writing down the wrong gain setting,

6. Using different integration times for the variable and comparison star and forgetting to apply the appropriate multiplicative factor,

7. Writing down the wrong filter,

8. Making a 1–hour, 100–day, or 1–year error in recording the time of observation.

Gross errors cannot and should not be treated statistically. Usually the only recourse is to recognize the gross error because of its extreme discordance relative to all the other observations and delete that observation altogether. Occasionally, however, you can detect the mistake itself (a scale reading error, a gain step error of exactly $0^m5$, an arithmetic error, an exact 1–year error in recording the time, etc.) and make the appropriate correction.

## 14.3 Sources of Error

### 14.3.1 Scintillation

As discussed in Chapter 9, scintillation produces a random error or noise in any deflection. This error cannot be avoided altogether and can be diminished somewhat only by taking longer deflections, using narrow bandpasses, restricting observing to small zenith distances, using a telescope of large aperture, and observing from a high altitude site.

### 14.3.2 Photon noise

Quantum mechanics tells us that (star) light arrives not as a continuous stream of luminous energy but rather as a train of discrete packets of energy called photons. The photons do not arrive at a perfectly uniform rate. Randomness, described by Poisson statistics, and clumping, described by Bose–Einstein statistics, lead to an inherent statistical uncertainty in any measure. If randomness were the only factor, then a signal which included $n$ photons or secondary electrons or pulses would have an uncertainty or statistical error of $\sqrt{n}$, which is called photon noise. Expressed as a fraction, the error would be $\sqrt{n}/n = 1/\sqrt{n}$. Now you can see why fainter stars, for which $n$ is small, have larger photon noise. For example, if $n$ = 100 for a 10–second deflection, the fractional error is $\pm$ 10% = $\pm$ $0^m1$; if $n$ = 1,000,000 for a 10–second deflection, the fractional error is only $\pm0.1\%$ = $\pm$ $0^m001$.

Since photon noise is greater for fainter stars, whereas scintillation noise does not depend on star brightness, there is a certain crossover magnitude for any given telescope, bandpass, etc. where scintillation noise equals photon noise. Young (1974) gives examples of crossover magnitudes for telescopes of various apertures. Photometry of stars fainter than the crossover magnitude is termed photon-noise-limited photometry.

### 14.3.3 Variable sky transparency

This usually is, and should be, the limiting factor in determining the final accuracy of differential photometry of rather bright stars. Let us consider the case where sky transparency is varying rather rapidly relative to the ~10 minutes required to make a single-filter intercomparison between variable and comparison. An example would be thin, irregular, rapidly moving clouds. If the transparency varies randomly between some maximum value and some minimum value only 95% as great, then in the worst case one of the two stars might appear 5% brighter or fainter than it should and the resulting differential magnitude thus would be in error by 5%. Or, in the best case, the transparency could vary in such a way that the average transparency during the two comparison star deflections was very close to the transparency during the variable deflection and the resulting differential magnitude thus would not be much in error. If the sky transparency varies randomly by the same amount but did so very slowly and gradually relative to the ~10 minutes required for an intercomparison, then the technique of bracketing would cancel the effect of variable transparency and the resulting differential magnitudes would not be much in error.

From this discussion you can see the benefit of moving rapidly between variable and comparison during differential photometry, particularly if the sky transparency is varying somewhat slowly (on a time scale of ~20 minutes). On such a night the observing sequence for multifilter differential photometry would require too much time and you would be wise to do one-filter photometry on that particular night.

On nights of variable transparency, due to thin clouds or haze, those same clouds or haze will increase the sky brightness also, especially if the moon is up. This enhanced sky brightness will be variable just as the sky transparency is variable and that will introduce additional error. We do not discuss this, however, because unless you are observing a star which is very faint relative to the sky background, the variable transparency generates much more error than does the variable sky brightness.

Errors due to variable sky transparency and variable sky brightness will be random errors, in fact, one of the clearest examples of such.

### 14.3.4 A variable comparison star

If the comparison star varies from its mean brightness by a certain fraction of a magnitude, say $0\overset{m}{.}03$, then the resulting differential magnitude of the variable will be in error by the same amount: $0\overset{m}{.}03$. Such an error will appear random or systematic depending on whether the comparison star is varying extremely rapidly or extremely slowly.

### 14.3.5 A faint background star

If a nearby star too faint to be seen by eye in the diaphragm is inadvertently included in either the star + sky deflection or the sky deflection, for either the variable or its comparison star, then the resulting differential magnitude of the variable will be in error. If the faint companion is, say, 5 magnitudes (i.e., 100 times) fainter than the variable (or the comparison), then the variable (or comparison) will appear 1.01 times or $0^m01$ too bright or too faint, depending on whether the faint companion was included in the star + sky deflection or in its sky deflection. This is a non-trivial problem because a star 5 magnitudes fainter than a 7th magnitude star will be 12th magnitude, not easily seen in a small telescope. The error will be systematic or random depending on whether the companion star is included every time (for example, if you are using an automatic sky offset) or only occasionally (for example, if you offset by moving the telescope). The former case is easier to deal with, because the brightness of the faint background star can be measured *ex post facto* and all of your (contaminated) measurements can be salvaged by appropriate arithmetic.

### 14.3.6 Not using the same diaphragm for both variable and comparison

If you violate the rule of using the same diaphragm for a given intercomparison between a variable and its comparison, you might, for example, be including 97% of the variable star image's total light in a small diaphragm but 99% of the comparison star image's total light in a larger diaphragm. In this example, the resulting differential magnitude for the variable would be too faint by 2% or $0^m02$, as you can see by referring to Eq. (9.2.2).

### 14.3.7 Variable voltage from the power supply

If the voltage from the power supply varies and the resultant sensitivity of the photomultiplier varies or drifts only gradually, then the technique of bracketing each variable deflection between two comparison star deflections will make the resulting differential magnitude immune to such drifting. If, on the other hand, the voltage oscillates rapidly relative to the ±10 minutes required for one bracketing, then differential photometry will be affected. A change of, say, 0.1% in voltage will produce a much larger, perhaps ±1% change in sensitivity, a ±1% change in the size of the deflections and a ± $0^m01$ change in the star's brightness. The resulting differential magnitudes of the variable will be affected a bit less than $0^m01$, because of bracketing, but of that same order.

### 14.3.8 Star not centered in the diaphragm

The sensitivity of any photocathode surface varies considerably across its surface. The Fabry lens is designed to have starlight illuminate a large area of the surface rather than be focused onto a small part. Thus, as the star moves around in the focal plane (wanders around in the diaphragm) slightly, the area illuminated remains pretty much the same. The overall response, therefore, is not extremely sensitive to centering in the diaphragm, but it can vary by about 1% from one side of a large diaphragm to the other. Sensitivity variations of 1% will, of course, produce corresponding errors of 1% or $0^m01$ in the resulting star brightness. This error in general will be a random error, but it will be systematic if for some reason you always position the variable left of center and the comparison star right of center. At some time every observer should make a series of trial deflections with an arbitrary star positioned various distances away from diaphragm center and thus construct a contour map of photometric sensitivity. Such a map will help you gauge how much error to expect from imprecise centering or inaccurate telescope tracking.

### 14.3.9   Automatic sky offset errors

Some photometers have a convenient feature which moves the diaphragm a fixed amount, so that the sky background reading can be made without having to move the telescope itself. Because the offset position is significantly off axis, any vignetting which may be present will cause the sky brightness to be registered fainter than it should be. The previous error source, related to centering in the diaphragm, would affect sky offset in a similar way, causing the sky brightness to be registered fainter or brighter than it should be if the Fabry lens is not doing its job perfectly. There are instances where the sky offset registered spuriously faint by ±20% (Kaitting 1984, Bisard and Osborn 1984). Although this will have an insignificant effect if the sky brightness is very small relative to a very bright star, it will have an enormous effect if a very faint star's brightness is small relative to the sky brightness.

### 14.3.10   Photomultiplier tube wobbling in its socket

Even if a star is always centered accurately in the diaphragm center, it might be that your photomultiplier tube wobbles in its socket or some other mechanical flexure problem causes the Fabry lens' illuminated beam to fall on different areas of the photocathode surface. The result will be similar to that caused by inaccurate centering. The error, in general, will be a random error, but it could be systematic if for some reason the tube always tips one way when the variable is observed and the other way when the comparison is observed.

### 14.3.11   Nonlinearity in photodetector response

As we said before, meaningful photoelectric photometry requires the photocell response to be linear, i.e., that voltage or current output be directly proportional to light intensity incident on the photo-sensitive surface. The various causes for such non-linearity were discussed in Chapter 4, but let us consider saturation, which can happen to any cell if too bright a star is observed, as a good example. Saturation sets in when the telescope aperture is sufficiently large, the voltage supplied to the photomultiplier sufficiently high, or the star sufficiently bright that your particular photomultiplier can no longer produce a current which is proportional to the intensity of the incoming light. For many photomultipliers, like the 1P21, significant saturation sets in beyond about 1 microamp. The particular 1P21 which was used with the Dyer Observatory 24–inch reflector for about 20 years, operated at −900 volts, generated about 1 microamp when observing a star of V = $3^m0$ with a wide-band visual filter. Each observer should, however, determine and remember the limit for his particular telescope and photomultiplier. If you observe a star about 1 magnitude brighter than your own limit, you can expect the resulting current to under-register that star's brightness by something like 1%. If by good fortune the variable and its comparison are very nearly the same brightness in a certain bandpass, then the resulting differential magnitude will be very nearly correct. But, if only one star is too bright, then that one will appear 1% or 0ͫ01 too faint. Such an error definitely will be a systematic error for a given variable-comparison pair.

### 14.3.12 Nonlinearity in the recording device

As we said before, meaningful photoelectric photometry requires that the response of the chart recorder (or V/F converter or pulse counter or other recording device) be linear also, i.e., that deflection length be directly proportional to incoming voltage over its entire full scale. Nonlinearity can take many forms but let us say that 5.00 millivolts input gives a deflection of exactly 50% full scale but that 10.00 millivolts input gives a deflection only 99% full scale. If, by good fortune, the variable and its comparison give deflections which are very nearly equal in length, no matter whether full scale or less, then the resulting differential magnitudes will be very nearly correct. But if the variable deflections are near full scale and the comparison star deflections are near half-scale, then in this example the variable will appear 1% or 0ͫ01 too faint. The resulting error can be systematic for a given variable star if the variable deflections were all near full scale and its comparison star deflections were all less than full scale.

### 14.3.13 Pulse coincidence or dead time error

This is a potential source of error only if you use a pulse counter, but that error can be incredibly large if you try to do photometry of too bright a star. Africano and Quigley (1977, Fig. 14–2) plot a nomogram which illustrates dramatically how huge an error (the magnitude difference corresponding to the ratio $N/n$) can result if the count rate is too fast. For $n = 4$ million photons/second, the error will be 0ͫ13, 0ͫ25, 0ͫ35, or 0ͫ48 depending on whether $\delta$ is 30, 50, 70, or 90 nanoseconds, respectively.

Our recommendation is the following. You should actually determine $\delta$ for your particular photon counter, not just accept the value provided by the manufacturer. You should always compute the dead time correction even if only to convince yourself that it is negligibly small. You should actually apply the correction if it is larger than about 0ͫ001. Finally, because the correction is only an approximation and hence, itself, becomes uncertain when it is large, you should distrust any magnitudes which require a dead time correction larger than about 0ͫ03, even if you use an improved correction equation such as given by Fernie (1976). For $\delta = 50$ nanoseconds this limits a 12.5–inch telescope to stars fainter than V = 4ͫ5. Current-measuring photometers are limited by charge buildup near the anode to currents which correspond typically to a star about 1 magnitude brighter.

### 14.3.14 Amplifier calibration error

When we use an amplifier to amplify a signal by a certain factor, it is obviously essential to know exactly what that factor was. The coarse gain settings on any operational amplifier (usually factors of 10 or, equivalently, 2ͫ5) rely on resistors of very high resistance and are almost never exactly equal to their stated value. The amplifier calibration determines the difference between the nominal factor and the actual factor. Typically, the difference is on the order of 1%. If you have neglected to calibrate your coarse gain steps, or if you did so long ago and they have changed significantly in the meantime, or if you did so at one temperature but observed at a considerably different temperature, then errors of around 1% or $0^{\text{m}}_{\cdot}01$ can be introduced. If the very same coarse gain setting is used for both variable and comparison, no error will be introduced. If, however, one is used for the variable and another for the comparison, but the same every time, then a systematic error will be made. If various different coarse settings are used from night to night and/or if observing is done under wide ranges of temperature, the errors will be more or less random.

### 14.3.15   Errors in measuring a deflection

Almost every observer who begins to measure his first grassy deflection questions his ability to estimate the average level accurately and impartially. Somewhat surprisingly, but fortunately, the measuring of deflections is not responsible for introducing much error over and above that inherent in the scintillation or photon noise. Experience has shown that, given the same grassy deflection, two different people will estimate the average level very nearly the same. The error in measuring a deflection typical of those most commonly encountered (with grass about $\frac{1}{4}$ inch high) is only a very few thousandths of a magnitude and often only about $0^{\text{m}}_{\cdot}001$. Moreover, such errors will be almost entirely random, not systematic.

### 14.3.16   Dusk and dawn errors

Some enthusiastic observers begin observing at dusk before the sky is sufficiently dark and continue observing when the sky has begun to brighten again at dawn. This can produce an error if only one sky deflection is taken, for example, consistently *after* each star deflection. Let us say the sky deflection at dusk, taken just after the star + sky deflection, is 9% of the star + sky deflection. If a sky deflection had been taken just before, it would have been 11%. Thus, a sky deflection of 10%, not 9%, should have been subtracted. This will make the net star deflection appear 1.1% or $0^{\text{m}}_{\cdot}012$ too bright. This error will to some extent cancel itself in the process of bracketing each variable deflection between two comparison star deflections, but only partly. Observers with an automatic star-to-sky switch can eliminate this error conveniently by taking the sky deflection always between two star deflections. All observers, however, can avoid the error altogether simply by waiting until the sky is sufficiently dark.

### 14.3.17   Errors in the assumed value of $k'$

As we explained before, differential photometry is inherently capable of far greater accuracy than absolute photometry because atmospheric extinction dims light from the variable and its comparison star by very nearly the same amount. Many observers do not appreciate how powerful this benefit is. To demonstrate this in a simple way, let us consider the effect of differential atmospheric extinction on two stars with the same right ascension, a variable at $\delta = +30°$ and a comparison star at $\delta = +29°$, observed from an observatory at latitude $\phi = +30°$. We will use the secant approximation to compute airmass for both stars with Eq. (9.1.2) for various values of the hour angle $h$. These values of $X$ and the resulting values of $\Delta X$ are given in Table 14–1. The correction for differential atmospheric extinction is the $k'\Delta X$ term in equations like (11.2.1), (11.2.2), or (11.2.3). Typically, the value of $k'$ is guessed or assumed for a given night and generally is within about $\pm$ 0$^{\text{m}}$1 of the actual value on that night. Therefore, the error in the differential extinction correction will be roughly 0.1 $\Delta X$. From Table 14–1, you can see that $\Delta X$ is less than 0.01 and the error in the differential extinction correction therefore is less than 0$^{\text{m}}$001, entirely insignificant, for observations obtained during the 8–hour interval 4 hours either side of the meridian. The situation would not be as favorable, however, if the angular separation was greater than 1°, the declination more southerly, or the assumed value of $k'$ in error by more than 0.1. Because the assumed value of $k'$ can be either larger or smaller than the actual value on a given night, and because variables observed at various times of the night and on many different nights can have $\Delta X$ values which are very different in size and both positive and negative, errors in the differential extinction correction will act predominantly as random errors in a long-term observing program.

| h | $X(\delta = +30°)$ | $X(\delta = +29°)$ | $\Delta X$ |
|---|---|---|---|
| **TABLE 14–1** | | | |
| **ILLUSTRATIVE EXAMPLE** | | | |
| **OF** | | | |
| **DIFFERENTIAL AIR MASS** | | | |
| −6:00 | 4.000000 | 4.125331 | −0.125331 |
| −5:00 | 2.251673 | 2.280785 | −0.029112 |
| −4:00 | 1.600000 | 1.609990 | −0.009990 |
| −3:00 | 1.281509 | 1.285351 | −0.003842 |
| −2:00 | 1.111705 | 1.113128 | −0.001423 |
| −1:00 | 1.026226 | 1.026654 | −0.000428 |
| 0:00 | 1.000000 | 1.000152 | −0.000152 |
| +1:00 | 1.026226 | 1.026654 | −0.000428 |
| +2:00 | 1.111705 | 1.113128 | −0.001423 |
| +3:00 | 1.281509 | 1.285351 | −0.003842 |
| +4:00 | 1.600000 | 1.609990 | −0.009990 |
| +5:00 | 2.251673 | 2.280785 | −0.029112 |
| +6:00 | 4.000000 | 4.125331 | −0.125331 |

### 14.3.18 Error in the assumed value of $k''$

The correction for the color-dependent part of atmospheric extinction is the $k''\overline{X}\Delta(\text{B–V})$ term in equations like (11.2.1), (11.2.2), or (11.2.3). Whereas $k'$ and $k''$ are sometimes referred to as the first-order and second-order extinction coefficients, respectively, this terminology can be misleading in differential photometry because the $k''$ term often produces a far greater correction that the $k'$ term. This is because only the $k'$ term contains $\Delta X$, which is so small in differential photometry, whereas the $k''$ term contains X, which is never less than 1.0.

We have advised using an assumed value of $k''$ for each bandpass, arguing that such a guess is probably reliable to about $\pm$ 0ᵐ01. Even if $k''$ is determined explicitly, it is unlikely to be more reliable than about $\pm$ 0ᵐ005. Using the $\pm$ 0ᵐ01 estimate we see that the $k''$ term, for a star pair differing by 0ᵐ3 in B–V and observed near the zenith ($X \approx 1$), will be uncertain by $\pm$ 0ᵐ003. Here you can see the importance of a close color match between a variable and its comparison. This error will be essentially systematic, i.e., constant, for a given variable-comparison pair provided all the observations were obtained at roughly the same airmass and the variable does not change randomly by a large amount in B–V as it varies.

### 14.3.19   Errors in the transformation coefficients

The differential transformation correction has the form $\epsilon\Delta$(B–V) or $\epsilon\Delta$(U–B) if we use an equation like (11.2.7), (11.2.8), or (11.2.9). In most cases, it is easy to determine your transformation coefficient to within $\pm0.01$, assuming there are no complications from nonlinear or multivalued transformation. Moreover, your particular value is quite likely to remain constant provided you do not alter your photometric equipment in any of the ways discussed in Chapter 10. Using the $\pm0.01$ estimate we see that, for a star pair differing by 0ᵐ3 in B–V, the differential transformation correction should be uncertain by about $\pm$ 0ᵐ003. Here also you can see the importance of a close color match between the variable and its comparison. This error is one of the best examples of a systematic error, since it will be quite constant for any one variable-comparison pair observed at any one observatory with the same photometric equipment. This is true, provided only that the B–V of the variable does not change randomly by a large amount as the variable varies.

### 14.3.20   Red leak

Just as a blue filter is "blue" because it transmits only in the blue part of the spectrum, so an ultraviolet filter should transmit only ultraviolet light. Unfortunately, many ultraviolet filters have a red leak which transmits a significant percentage of red light also. The situation is helped somewhat, but not entirely, by the fact that many commonly used photomultipliers are much less sensitive in the red than in the ultraviolet; and the popularity of newer photomultipliers with extended red response means the red leak problem can be more serious than before. A good example is the Corning No. 9863 filter, which unfortunately was used originally to define the U bandpass in the UBV system. It has about 50% transmission at about 7000 angstroms, where the photocathode of a 1P21 photomultiplier (also used to set up the UBV system) still has appreciable though diminished sensitivity. You can easily see the red leak problem. In trying to determine U magnitudes of a very red star, which itself emits extremely little ultraviolet light, most of the "U deflection" is produced not by ultraviolet light but rather by red light coming through the red leak. Without appropriate correction, these deflections would be meaningless for deriving U magnitudes; and even with appropriate correction the ultraviolet light might be so feeble compared to the red light which must be subtracted that no U magnitudes of useful accuracy could be obtained.

There have been basically three approaches to solve the red leak problem.

1. Select an ultraviolet filter which has little or no red leak. An excellent choice is 1 mm or 2 mm of Schott UG–2, which has only about 1% the red leak of Corning No. 9863. Most observatories today doing UBV photometry use this filter and we recommend it. An earlier approach, suggested by Kron (Wood 1963), was to "plug the leak" with a cell of carefully prepared and frequently changed copper sulfate solution. But this approach is not recommended anymore.

2. Measure the red leak itself by taking a second deflection through a combination of the same ultraviolet filter and another filter which absorbs the ultraviolet light while transmitting the red light. Then subtracting the second deflection from the first should give a deflection proportional to the ultraviolet light alone. This approach, discussed in detail by Shao and Young (1965), is quite straightforward and we recommend it if, for some reason, you are required to use a leaky filter like Corning No. 9863 or use a red-sensitive tube for ultraviolet photometry.

3. As a variation of the above approach, you can measure the red leak for stars of various color index and use these measures to determine a correction curve. Thus, for a star of a certain B–V, the curve gives you a correction to convert your observed U magnitudes into uncontaminated U magnitudes. Shao and Young (1965) discuss the several drawbacks in this approach.

The red leak in ultraviolet filters is not the only one. Blue filters have red leaks also and photometry with narrow band filters is vulnerable to even a slight leak in the large continuum spectral regions they are supposed to exclude.

### 14.3.21  Dome errors

If inattentiveness allows your telescope to be occulted partially by one edge of your dome, that, of course, will introduce error. Fortunately, the technique of differential photometry will minimize such errors but, because the variable and its comparison are always separated somewhat in the sky, there will be some residual error nevertheless. This is an error that should be avoided rather than discussed.

### 14.3.22  Focusing errors

Depending on how the optics in your photometer are arranged, it is possible that the star will be in focus in the viewing eyepiece, but the light may not be focused at the diaphragm. The result would be a non-square beam scan, something very difficult to unravel if you do not observe with a strip chart recorder. Before you figure this problem out, it appears that the telescope cannot guide well at all, even with a large diaphragm, and that the photocathode has a "hot spot" where the response seems to have a sharp maximum.

### 14.3.23  Other errors

Although our list is already quite long, we cannot pretend to have considered all possible sources of error. As the observer, you should keep your eyes open at all times. In the course of your productive years, you will probably discover a new source of error yourself. Trouble can strike from any direction. Young (1974) discusses polarization, magnetic, and gravitational effects. We haven't discussed light leaking into the photometer, something which can happen but should be avoided. We, ourselves, have experienced trouble when a loose dark slide slipped in gradually throughout the night as the angle of the telescope changed. We have also experienced trouble with an open-tube telescope as light from the chart recorder found its way into the photometer head, by variable amounts depending on exactly where the observer was standing in front of the charter recorder while making notes. We have even heard of an unfortunate photoelectric photometrist in Texas who had problems with flying ants which somehow entered his photometer head and, throughout the entire night, made random walks across his Fabry lens. More recently, we have heard another had a similar experience with an earwig.

## 14.4  Determining Standard Devitation ($\sigma_1$) and Mean Error ($\sigma_m$)

The photoelectric photometrist should, as any experimental scientist should, become familiar with the basic aspects of probability and statistics, especially Gaussian statistics, which treats the occurrence of small random errors.

From the three $\Delta m$ values resulting from the basic sequence for differential photometry you should compute the mean or average ($\overline{\Delta m}$) and then the standard deviation ($\sigma_1$) of each value from the mean. The formula for computing standard deviation is $\Sigma\left(x - \overline{x}\right)^2 = \left(x_1 - \overline{x}\right)^2 + \left(x_2 - \overline{x}\right)^2 + \left(x_3 - \overline{x}\right)^2$

$$\sigma_1 = \sqrt{\frac{\Sigma(x - \overline{x})^2}{n-1}}, \qquad (14.4.1)$$

*Raw*
*Inst*
*Mag diff*

where $x$ stands for each measured value (in this case $\Delta m$), $\overline{x}$ is the mean or average of the three (in this case $\Delta m$), and $n$ is the number of measurements (in this case $n = 3$). In the numerical example given in Table 13–4,

$$\overline{\Delta V} = +1\overset{m}{.}724$$
$$\sigma_1 = \pm 0.014 \quad \text{in} \quad V,$$
$$\overline{\Delta B} = +1\overset{m}{.}502$$
$$\sigma_1 = \pm 0\overset{m}{.}012 \quad \text{in} \quad B.$$

*1.740*
*1.713*
*1.720*
*3/5.173*
*1.724*

You should aim for values of $\sigma_1$ around $\pm 0\overset{m}{.}01$. With reliable and carefully calibrated equipment, optimum observing technique, favorable observing conditions, sufficiently bright stars, etc., a good photoelectric photometrist sometimes can achieve a standard deviation around $\sigma_1 = \pm 0\overset{m}{.}003$ in differential photometry. Differential photometry with a standard deviation as large as $\pm 0\overset{m}{.}03$ can still be useful, if the variable star has a light curve of sufficiently large amplitude, for example; but the observer should not be satisfied unless unavoidable sky transparency variations are the limiting factor.

The value of $\sigma_1$ indicates how reliable each value of $\Delta m$ is. Obtaining more than three values of $\Delta m$ will not change $\sigma_1$. The uncertainty or error of the mean itself, however, does diminish as $n$ increases. The formula for computing mean error is

$$\sigma_m = \frac{\sigma_1}{\sqrt{n}} = \sqrt{\frac{\Sigma(x - \overline{x})^2}{n(n-1)}}, \qquad (14.4.2)$$

where the symbols are the same as in Eq. (14.4.1). In the numerical example given in Table 13–4, $\sigma_m = \pm 0^\text{m}008$ for V and $\sigma_m = \pm 0^\text{m}007$ for B.

Both $\sigma_1$ and $\sigma_m$ are estimates of the internal error, because they have been deduced from the data themselves without recourse to any external bases of comparison. They do not and logically cannot provide any indication whatsoever of any systematic error which may be inherent in the data. Since systematic errors can (and, in fact, usually do) occur, internal errors are optimistic in that they underestimate the actual errors.

If mean errors computed from standard deviations do describe the actual errors, i.e., if there are no systematic errors, the Gaussian statistics tell us what quantities like $\sigma_1$ and $\sigma_m$ mean. Again, we use the numerical example in Table 13–4. The standard deviation of $\sigma_1 = \pm 0^\text{m}014$ tells us there is a 65% probability that any one value of $\Delta V$ will differ from the mean of $\overline{\Delta V} = 1^\text{m}724$ by less than $\sigma_1 = 0.014$, a 95% probability that any one value will differ by less than $2\sigma_1 = 0^\text{m}028$, and a 99.5% probability that any one value will differ by less than $3\sigma_1 = 0^\text{m}042$. Similarly, the mean error of $\pm 0^\text{m}008$ tells us that there is a 65% probability that the mean of $\overline{\Delta V} = 1^\text{m}724$ will differ from the true or correct value of $\Delta V$ by less than $\sigma_m = 0^\text{m}008$, a 95% probability it will differ by less than $2\sigma_m = 0^\text{m}016$, etc.

You can see in Eq. (14.4.3) that $\sigma_m$ can be made smaller by increasing $n$; i.e., by making more intercomparisons between the variable and its comparison star. But the effort may not be worthwhile because of the square root sign in Eq. (14.4.3). If the photometry of observer A usually shows $\sigma_1 = \pm 0^\text{m}01$ whereas that of observer B usually shows $\sigma_1 = 0^\text{m}03$, the corresponding values of $\sigma_m$ will be in the same proportion providing $n$ is the same for both A and B. If observer B wishes to match the smaller $\sigma_m$ of observer A by increasing $n$, he must obtain 9 times as many inter-comparisons, 27 instead of 3! His effort might be spent better trying to learn why his $\sigma_1$ is so large. Moreover, it must not be forgotten that increasing $n$ has absolutely no effect on any systematic error which may be present.

Gaussian statistics tells us that statistical weight is inversely proportional to the square of the mean error:

$$w = \frac{1}{\sigma_m^2}. \tag{14.4.3}$$

For example, if a campaign coordinator wanted to average the value $\overline{\Delta V} = 1^\text{m}724 \pm 0^\text{m}014$ in Table 13–4 with the value $\Delta V = 1^\text{m}732 \pm 0^\text{m}007$ provided by another observer, he would assign a weight of $w = 1$ to the first and $w = 4$ to the second. The resulting weighted average would be

$$\frac{1(1^\text{m}724) + 4(1^\text{m}732)}{5} = 1^\text{m}730. \tag{14.4.4}$$

Eq. (14.4.3) should be used to derive weights only from true or actual mean errors. Thus, $\sigma_m$ derived from Eq. (14.4.3) would be appropriate only if you somehow had assurance there was no appreciable systematic error in the mean.

Gaussian statistics tell us one final thing, which can be very useful in our attempt to achieve maximum accuracy. If more than one root cause is contributing to our overall accidental error, as is usually the case, then their contributions combine, not simply additively, but rather by the square root of the sum of the squares. The total standard deviation $\sigma_T$ is given by

$$\sigma_T = \sqrt{\sigma_A^2 + \sigma_B^2 + \sigma_C^2 + \cdots},\qquad(14.4.5)$$

where $\sigma_A$, $\sigma_B$, etc. represent the individual contributions if they were acting alone. Let us say scintillation noise is $\sigma_A = \pm 0^m01$, photon noise is $\sigma_B = \pm 0^m01$, and variable sky transparency is responsible for $\sigma_C = \pm 0^m03$. The total would be $\sigma_T = \pm 0^m033$. Somehow eliminating both $\sigma_A$ and $\sigma_B$ would lower $\sigma_T$ to $\pm 0^m030$, hardly any improvement. Eliminating just $\sigma_C$, however, would lower $\sigma_T$ to $\pm 0^m014$, which corresponds to more than a five-fold increase in statistical weight. The message is clear. Ascertain which of the many contributory sources of error is the *largest* and work first to minimize that one. Effort spent eliminating lesser errors will be effort wasted.

# Chapter 15

# KEY FACTORS IN THE CHOICE OF A PROGRAM

In this chapter we want to discuss factors you should consider when choosing an observing program for yourself. You should do what you can do well at your own observatory. Choose a project which seems "just right" for your particular observatory and its photometric equipment. Don't be heroic and pick stars which are too faint, too far south, etc. Stars you can't handle can be handled at other observatories. Feel free to acknowledge your personal limitations, such as available time, stamina, or money. Choose a project which for you is not exhausting personally or financially. Demand that the observing you do be scientifically worthwhile. Obtaining good observations of a variable which is entertaining but frankly does not need repeated routine observing can be compared to "collecting more seashells." To a large extent, however, you often will have to rely on the judgment of someone else in selecting a scientifically worthwhile observing program, and there is no foolproof way to know at the outset how wise that judgment is.

One useful criterion, though, is publication. Good observations of important variables which need observing can and should be published in reputable scientific journals. Astronomical journals which publish articles periodically represent the arena in which current scientific debate occurs and scientific progress is made. Therefore, if you find your observations are good in quality but are not being published or utilized in some other substantial way, it is appropriate for you to feel that your efforts are not being made use of as effectively as they should be, and you should consider changing your observing program. A regular feature in the *I.A.P.P.P. Communications* lists papers co-authored by amateurs in regular astronomical publications. The total to date (at least 150 papers in 20 different publications, as of the end of 1985) attests to the fact that scientifically worthwhile research can be accomplished by amateur photoelectric photometry and is made accessible to the astronomical community.

## 15.1 Long-Period (Slow) Variables

Variables which vary slowly and gradually require only one observation during each night; observing them continuously throughout an entire night will not tell you much more. The recommended basic sequence described in Section 4 of Chapter

12 is perfect for them. Such variables are ideal for you if your observatory site often has nights which are clear for a couple of hours but seldom has nights which are clear from dusk to dawn.

Long-period variables are often observed profitably as part of a cooperative campaign in which observers at various geographical locations participate, since not all will be clouded out at the same time and fairly complete coverage night by night can result from the collective effort. There is another advantage to observing slow variables. If multifilter photometry is needed but your sky conditions require you to go through the recommended basic sequence separately in each filter, this is no drawback since the slow variables will not have changed much in a half hour or so.

## 15.2    Short-Period (Fast) Variables

A good example of a fast variable is a short-period eclipsing binary like RZ Cassiopeiae, which undergoes rather deep eclipses lasting only a few hours. To observe such a variable effectively you will need a night of relatively good photometric quality so you can observe the entire night and also so you can get more than one deflection on the variable between comparison star deflections.

With fast variables it is best to do one-filter photometry, because it can be done more quickly and more accurately than with multifilter photometry and because time of minimum or maximum brightness is not, in general, a function of wavelength.

Another requirement is to record your time accurately, by synchronizing your clocks to the nearest couple of seconds with WWV and marking your times often enough and precisely enough every time a variable deflection is made. Also, with fast variables it is necessary to compute the heliocentric correction with sufficient precision, namely, to the nearest $0\overset{d}{.}0001$.

## 15.3    Extremely Rapid Variables

In this book we say very little about extremely high speed photoelectric photometry because the instrumentation needed for it can be quite expensive and sophisticated and the techniques required are quite specialized.

Examples of extremely rapid variables would be the flare stars like UV Ceti, which undergo solar-type flares for which the rise time can be as short as a minute, and the white dwarf in the old nova DQ Herculis, which pulsates with a period of 71 seconds, the dwarf novae like U Geminorum and SS Cygni which flicker on time scales as short as 1 second, and the optical pulsar in the Crab Nebula, which flashes briefly every 0.03 seconds. Various occultation events also require high speed techniques, for example, occultations of stars by planets, asteroids, or the moon.

It can help in some cases to use a strip chart recorder which can drive the paper at least ten times faster than the usual inch per minute rate. Best of all, however, are systems in which output from the photomultiplier is either recorded directly on magnetic tape, which can be made to move very rapidly, or is stored directly in a computer's memory. In all these cases, it is advantageous to have time signals, from WWV for example, automatically registered directly on the record, either as a tick on chart paper or a blip on magnetic tape.

## 15.4 When Wavelength is Not Important

One can do differential photoelectric photometry more easily, more quickly, and more accurately with one filter than with several filters. Therefore, unless multifilter photometry is needed for a good reason, it is best to do single-filter photometry whenever you can.

Single-filter photometry is best for determining times of maximum or minimum light for objects in which the light varies as the geometry changes. Examples would be eclipsing binary stars, lunar occultations, and tumbling asteroids.

Single-filter photometry is also well suited for the BY Draconis-type and RS Canum Venaticorum-type variables, which we discuss in more detail in the next chapter. These stars vary in light as an asymmetrically spotted star rotates on its axis, turning a spot group alternately into and out of view. These are dark spots, roughly $1000°K$ cooler than the surrounding photosphere, which is around $4000°K$. In principle, the light variations should be accompanied by color changes since a cooler photosphere does emit its own light and that light would be redder. In actuality, however, because luminous flux is proportional to the *fourth* power of the temperature above absolute zero and because the spot group rarely covers more than about 25% of one hemisphere, light contributed by the cooler region is so feeble that it changes the color of the starlight very little, often too little to be detected. One has to observe in the far ultraviolet or in the near infrared before the color change in these variables becomes important.

If the object of your observing program is to determine times of maximum or minimum light specifically in order to ferret out the period, then there is no reason at all to do multifilter photometry. In fact, you could profitably do no-filter or wide-open photometry in this instance if the star happened to be uncomfortably faint for your telescope.

## 15.5 When Wavelength is Important

A good example of variables in which the choice of wavelength is important is those eclipsing binaries consisting of two stars which are very different in temperature. In eclipsing binaries with circular orbits, the amount of projected surface area eclipsed at the two conjunctions is the same. Therefore, the deeper (primary) eclipse will be the eclipse of whichever star has the greater surface brightness, i.e., emits more luminous flux per unit area. That star will be the hotter one. Conversely, the shallower (secondary) eclipse will be the eclipse of the cooler star. As you go to progressively longer wavelengths, secondary eclipse becomes deeper while primary eclipse becomes shallower. Thus for many eclipsing binaries (the Algol-type eclipsing binaries with their characteristically deep primary minima being the best example), secondary eclipse is virtually undetectable in the ultraviolet, is detectable but quite shallow in the visual, and is quite apparent in the red and infrared. For Algol, itself, (Soderhjelm 1980) the depth of secondary eclipse ranges from $0^{\text{m}}02$ in the far UV, to $0^{\text{m}}3$ in $N$. As you go to progressively shorter wavelength the opposite happens: primary eclipse becomes deeper while secondary becomes shallower. For V471 Tauri, an important eclipsing binary containing a white dwarf and a red dwarf, primary eclipse is $0^m04$ in V, $0^m09$ in B, and $0^m36$ in U (Nelson and Young 1970). If the two components of an eclipsing binary are very similar in temperature, then of course the depths of primary and secondary eclipse will be comparable in depth and will change very little as a function of wavelength.

There are other examples where wavelength is important. Flares, in the flare stars like UV Ceti, are always much greater at shorter wavelengths, especially U. The amplitudes of Mira-type variables are very different at different wavelengths, typically 6 or 7 magnitudes at V but only 1 or 2 magnitudes at I. At minimum brightness, however, the stars themselves are many magnitudes brighter in R and I than they are in B or V (Wing and Hall 1983). Although spotted variables like the RS CVn binaries show almost identical light curves in B and V, i.e., they show very little color change as they vary in brightness, it is possible to determine the temperature of the dark regions if one contrasts the light curves at R and I with the light curves at B and V (Eaton 1983). And, although tumbling asteroids show virtually identical light curves at all wavelengths, the color index itself can be useful in saying something about an asteroid's chemical composition (Smith 1985).

If you want to do photometry of a star which is almost too faint for your telescope but you wish to be observing in the bandpass of some standard photometric system, then you should choose whichever bandpass gives you the strongest signal. In general, this would be a long-wavelength bandpass for a cool star and a short-wavelength bandpass for a hot star, but you must also take into account the fact that any given photocell will have its greatest sensitivity at some wavelength region. The 1P21 photomultiplier, for example, is most sensitive in the blue, whereas the photodiode in the Optec photometric unit is most sensitive in the red.

## 15.6     Solid Coverage Versus Intermittent Coverage

If your observatory site gives you relatively few clear nights which remain clear from dusk to dawn or if other factors allow you to observe only occasionally, then you must choose variables for which this limitation is not very important and intermittent photometric coverage can be useful.

One possible good choice for intermittent observing would be a periodic variable which repeats exactly the same light variation every cycle. In this case, you could cover whatever portion of its light curve is available whenever your sky is clear and then, after observing for many (widely separated) nights, combine all of your observations onto the same light versus phase plot to produce the complete light curve.

Another possible good choice would be a variable with a smooth light curve and such a long period that an interruption of one week or so would not leave a significant gap in the light curve. Mira variables, with smooth light curves and periods around a year long, are perfect in this respect.

Variables with smooth light curves and periods longer than one or two weeks are attractive in that you need observe them only once during the night; repeated observations during the same night would appear more or less as one point on the light curve.

There is a problem, though, with variables which do not have periods so long that interruptions of a week or so can be tolerated and which are not periodic or do not repeat exactly the same light curve each cycle. Any one observer at a relatively poor observing site can get no more than a fragmentary light curve of such variables, often too fragmentary to be of much scientific use. If, however, he is participating in a cooperative campaign in which a half dozen or dozen other observers work on the same star whenever their sky conditions permit, then the result of the collective effort can be very nearly continuous coverage and a very nice light curve. Two such examples are shown in Fig. 16–1.

Solid photometric coverage of a variable, i.e., deflections taken more or less continuously throughout the entire night without interruption, is best done at observing sites where there are many nights during which the photometric quality is good and remains so the entire night. Understandably, the big national observatories, like Kitt Peak National Observatory and Cerro-Tololo Inter-American Observatory, and the observatories at large universities, like Lick Observatory for the University of California, are situated at such sites. Most astronomers using these facilities do not live there permanently but rather travel there from several hundred miles away after having submitted an application for a certain block of observing time on a particular telescope. Since these blocks range from one or two days to a few weeks and rarely are longer than a month, these observers are at a fundamental disadvantage for observing variables with periods longer than a few weeks. If your backyard telescope is at a site which is not favorable for continuous coverage of variable stars, do not despair; you have a fundamental advantage precisely because your observatory is in your backyard. By engaging in intermittent photometry but doing so persistently month after month, you can handle long-period variables which so many large observatories simply cannot.

## 15.7 When Your Particular Geographical Longitude and Latitude Can Make You Valuable

At some time you may discover that your particular observatory is unique and can provide photometry which no other observatory can, simply by being located where it is, at a particular geographical longitude and latitude.

The most obvious example of this is certain occultation phenomena. Examples would be grazing lunar occultations of bright stars, occultations of stars by asteroids, rare occultations of bright stars by planets, and the many occultation and shadow phenomena which occur in the complex planet-satellite systems of Jupiter and Saturn (Asknes and Franklin 1985). In the extreme case, grazing lunar occultations, tens of meters in geographical location can make an important difference. Another example would be an international campaign for the continuous photometric monitoring of an important long-period variable star throughout one complete orbital cycle. A perfect example of this was the international program of coordinated photometric and spectrographic observations of the infamous $\beta$ Lyrae continuously over a 35–day interval in 1959, almost three complete orbital cycles. This campaign, sponsored by Commission 42 of the International Astronomical Union, was coordinated by Dr. Gunnar Larsson-Leander, and the results have been published (Larsson-Leander 1969, 1970). Given that a certain star is in the right part of the sky for observing during a particular season, the only way to obtain 24–hour photometric coverage of it is to enlist the help of observatories distributed over the 360° of terrestrial longitude. Observatories situated at longitudes where there are not many observatories, islands in the Pacific Ocean, for example, become especially important in such a campaign. Such campaigns are organized to study important long-period variables, the light curves of which change significantly from one cycle to the next.

Somewhat similar to the above example is the problem posed by an eclipsing binary or other variable which has a period very nearly an integral number of days. An example would be Z Herculis with its orbital period of 3.996. Any one observatory, even by observing Z Herculis from dusk to dawn every night, could cover only four segments of the full light curve. The four gaps between would not become accessible to that observatory, even after a season had passed, because the period is so nearly exactly 4 days. Actually, it is the period expressed in sidereal days which is relevant in this context but, multiplying 3.996 by the fraction $366\overset{d}{.}2422/365\overset{d}{.}2422 = 1.002738$, we get $4\overset{d}{.}007$ sidereal days, which is still very close to an integral number. A complete light curve of Z Herculis can be obtained within one observing season only if photometry is obtained at observatories located at other terrestrial longitudes. Z Herculis is not the only example. Many important variables are afflicted by this same malady, having a period very nearly an integral number of sidereal days.

Your observatory can be important whenever a certain star needs night-by-night photometry during a certain interval of time. The reason for this is very simple. On any particular night, many observatories on the earth's surface will be unable to function because of cloud cover, but it is highly improbable that cloud cover will blanket the entire earth at any one time. Thus, your observatory very well could be one of the few which can function on a certain night when so many others cannot. We can think of several examples which illustrate this point.

The various satellite observatories orbiting the earth at any moment typically are scheduled to point at selected individual stars for certain intervals of time. Often the satellite observations (usually in the far ultraviolet or X–ray regions of the spectrum) require simultaneous ground-based photometry in bandpasses of the visible spectral window in order to be analyzed and interpreted in the most meaningful way. In a similar way, observations of stars by radio telescopes very often need simultaneous ground-based photometry to make those radio data as meaningful as possible. The paper by Weiler et al. (1978) is a good example of a campaign in which ground-based UBV observations by amateur and professional astronomers provided indispensable photometric backup for (1) far-ultraviolet observations by the Copernicus (OAO–III) satellite, (2) observations at centimeter wavelengths by radio telescopes at Green Bank, West Virginia and at Jodrell Bank in Manchester, England, (3) polarimetric observations made in Finland, and (4) spectroscopic observations made in the U.S. and Canada.

Another example would be those rare moments when an unanticipated celestial event occurs: a supernova outburst, a nova outburst, or a large radio outburst of some star. In such fleeting moments, hour-by-hour photometry, from whatever observatory has no cloud cover overhead, becomes priceless. A half-dozen photoelectric measures of the now famous Nova Cygni 1975 were obtained during the critical two days before the nova reached its maximum brightness. The giant radio outburst of V711 Tauri = HR 1099, which began late in February 1978, was observed faithfully by simultaneous photoelectric photometry on 14 of the first 18 nights during which the radio outburst occurred (Chambliss et al. 1978). The critical radio-spectroscopic-photometric results of this significant event were published together in the December 1978 issue of the *Astronomical Journal*, with several amateurs co-authoring the papers on photometry.

# Chapter 16

# HOW TO SELECT AN OBSERVING PROGRAM

We decided deliberately not to let this chapter be simply a list of recommended observing programs. There is a good reason for this. In the ongoing battle to expand the circle of knowledge and push back the frontier of ignorance, the battle lines are rapidly and constantly changing. For your efforts to be well spent, they should be directed against a specific well-defined target. Otherwise, you might find yourself just collecting more seashells. If we tried to list specific well-defined variable star problems suitable for differential photoelectric photometry and scientifically worthwhile at the time we write this book, then by the time you are reading the book those problems might be as relevant as last week's weather forecast. You should, however, judge the suitability of any variable star observing program by considering the criteria explained in Chapter 15.

There has, however, been much excellent material written about variable stars and this material can provide valuable background for anyone embarking on a photometric observing program. Comprehensive books reviewing variable stars in general include those of Merrill (1938), Campbell and Jacchia (1941), Glasby (1969), Hoffmeister (1970), Strohmeier (1972), Petit (1982), Tsesevich (1970), and Hoffmeister, Richter, and Wenzel (1985). The chapter by Abt in the book of Wood (1963) also belongs to such a list. Other books have concentrated on certain types of variables. These include the long-period variables (Campbell 1955), novae (Payne-Gaposchkin 1964), dwarf novae (Glasby 1970), supernovae (Clark and Stephenson 1977), nebular variables (Glasby 1974), RR Lyrae variables (Tsesevich 1969), flare stars (Gurzadyan 1980), pulsating stars (Kukarkin 1970), and binary stars (Batten 1973, Sahade and Wood 1978, Kopal 1978, 1979, Eggleton and Pringle 1985). Still other books, too numerous to list here, are edited collections of chapters or contributions written by various astronomers, each discussing one aspect with which he has some expertise. Such works often are published proceedings of an international colloquium or symposium convened to focus attention on one type of variable star or one specific problem in variable star research. Many of these are sponsored by the International Astronomical Union and as such are designated I.A.U. Symposia or Colloquia. I.A.U. Symposia bearing numbers 3, 51, 59, 67, 70, 73, 83, 88, 98, and 99 and I.A.U. Colloquia bearing numbers 4, 6, 15, 21, 29, 32, 42, 46, 53, 59, 65, 69, 70, 71, 72, 82, 92, 93, 103, 104, and 107 can be considered to have involved variable stars if we include the eclipsing binaries as variables. In addition, the Remeis Observatory in Bamberg, West Germany and the Konkoly Observatory in Budapest, Hungary, have been taking turns hosting the Variable Star Colloquia approximately every three years. By now there have been nine: in 1959, 1962, 1965, 1968, 1971, 1975, 1977, 1981, and 1986; the last seven of which were also I.A.U. Colloquia. The definitive source of information on current variable star research, of course, would be the report of Commission 27 (Variable Stars) to the International Astronomical Union, which convenes every three years and publishes the Commission Reports in the Transactions of the I.A.U. More timely are two periodical publications dealing specifically with variable stars: *Variable Stars*, published in the U.S.S.R., and the *Information Bulletin on Variable Stars*, published in Hungary. We conclude this already long paragraph by noting that section 9.2 of Hoffmeister, Richter, and Wenzel (1985) is an excellent, even more complete listing of general sources, compilations, and review articles pertaining to variable stars.

A symposium was held recently at the University of Toronto (Percy 1986), the topic of which was "The Study of Variable Stars Using Small Telescopes." The message was that small telescopes have been, still are, and will continue to be indispensable for studying variable stars. Not surprisingly, photoelectric photometry played a major role. The published proceedings of that symposium constitute be a valuable reference, for those with a small telescope equipped for photoelectric photometry, in connection with the selection of suitable observing projects.

## 16.1  Observing Programs Suggested by Another Astronomer

Your best bet is to rely on the advice of a competent astronomer who is actively involved in variable star research and whose publication record indicates work on scientifically worthwhile problems in the past. This advice can come from personal interaction with an individual astronomer or from reading recent articles or review papers, which quite often point to specific important problems unsolved at that time. It can be advantageous logistically to work closely with an astronomer at the university or observatory which happens to be located in your city or near your observatory, but this is not always possible and is not necessary. Lines, Lines, Boyd, and Genet (1985) discuss how this process can work in practice.

## 16.2  I.A.P.P.P.

There is an organization designed expressly to put small observatories capable of doing photoelectric photometry in contact with astronomers who know of timely research problems involving photometry which could be handled by small photoelectric observatories. This organization is called International Amateur–Professional Photoelectric Photometry (I.A.P.P.P.) and distributes the *I.A.P.P.P. Communications* to all observatories interested in either aspect of this endeavor. The address for I.A.P.P.P. is given in Appendix B. In the *Communications* professional astronomers can outline scientifically worthwhile problems suitable for photoelectric photometry with telescopes of small or modest aperture. An astronomer at any amateur or small college observatory wishing to take on such a problem then contacts the professional and they work together from that point onward. No photoelectric photometry data or results as such are published in the *I.A.P.P.P. Communications*, but the editors do provide advice, constructive criticism, and encouragement to those wishing to publish their results in reputable astronomical journals.

After more than five years since the founding of I.A.P.P.P. and after 22 issues of the *Communications*, a large number of possible observing projects have been described. They are summarized in Table 16–1, by category, with a reference to the issue of the *Communications* in which they appeared. Although variable star projects are in the majority, the diversity is wide. Target objects as small as asteroids and as large as galaxies are included. High-speed (millisecond) photometry is described, as well as suggested one-time observations of non-variable sources. And, as a response to those who fear their skies are so poor that they cannot do scientifically meaningful work, there are projects involving photoelectric photometry of sky brightness, to provide quantitative measure of light pollution. Relatively few of these projects have resulted in what would be considered sufficient photoelectric photometry, so it seems there are still more stars than astronomers.

## 16.3 Times of Minimum

Determining times of minimum light or times of mid-eclipse for eclipsing binaries is an activity which many small observatories enjoy and have engaged in profitably for a long time. Since before the turn of the century astronomers have made visual and photographic estimates of eclipsing binaries in order to determine times of minimum light. Such timings can be quite useful for many eclipsing binaries, one reason being that they were the only timings available, sometimes for 75–year intervals of time. There is an excellent discussion by Mallama (1974a, 1974b) of the accuracy which can be expected for timings based on visual estimates. There is no doubt, though, that photoelectrically determined times of minimum are especially valuable because of the far superior accuracy they are capable of achieving.

We feel it is safe to suggest eclipse timings as a worthwhile activity for differential photoelectric photometry because there is every indication that they will continue to be scientifically useful for some time to come. They are important in two ways. The first is simply to determine or improve the ephemeris so that future eclipses can be predicted with sufficient accuracy to be helpful in planning observing programs or so that astronomers making spectroscopic, satellite, radio, or polarimetric observations during eclipse can know which eclipse phases correspond to the times of their observations. The second is the study of period variations in eclipsing binaries. Probably around half of all known eclipsing binaries have orbital periods which vary enough to be detected after a few decades of observation. Although these changes complicate the problem of determining ephemerides which are useful for predicting future eclipses, the real interest is in the changes themselves. The problem has been attacked vigorously and competently, ever more so in recent years, but still remains essentially unsolved, one

| TABLE 16–1 |
| :---: |
| PHOTOELECTRIC OBSERVING PROGRAMS |
| DESCRIBED IN THE I.A.P.P.P. COMMUNICATIONS |

| Category | I.A.P.P.P. Comm. No. |
| :--- | :--- |
| I. Variable Stars | |
| Mira-type | 6, 14 |
| RV Tau-type | 4 |
| Cepheids & RR Lyr-type | 5, 6 |
| Semi Regular | 19 |
| Symbiotic | 17 |
| Be stars | 4 |
| T Tau-type | 21 |
| Eclipsing binaries | 7, 19 |
| RS CVn-type binaries | 3, 4, 9, 13, 14, 20, 21 |
| V1010 Oph-type binaries | 16 |
| W Ser-type binaries | 16 |
| Individual | 5, 6, 9, 11, 13, 16, 17 |
| | 19, 20, 21, 22 |
| II. Occultations | |
| Stars by asteroids | 14 |
| Stars by the moon | 10 |
| Galilean satellites | 19 |
| Pluto by Charon | 22 |
| III. Comets | 8, 9, 17 |
| IV. Asteroids | 5, 7, 11, 13, 14, 15, 16, 17, 21 |
| V. Moon | 14 |
| VI. Comparison Stars | 21 |
| VII. Galaxies | 3 |
| VIII. Sky | |
| Light Pollution | 9, 10 |
| Volcanic eruptions | 13 |

of the most serious and embarrassing unsolved problems in binary star research today (Sahade and Wood 1978).

Various associations of amateur variable star observers for many years have been organizing their members to make visual timings of eclipsing binary minima and publishing the results. Some of these include the A.A.V.S.O., the B.B.S.A.G., and the B.A.V. Addresses for these organizations are given in Appendix B. Rather recently some of these, the first two for example, have begun to organize photo-electric timings as well. These groups can be very helpful in suggesting particular stars to observe, providing finding charts, offering to reduce the data in a system-atic way, and publishing the resulting times of minimum. One potential drawback to watch out for is that such long-standing lists of program stars might not be culled or updated very often. Thus, they might contain stars which frankly do not need times of minimum nearly as much as others which are not on the lists. Here is where a researcher active in the field sometimes can make valuable recom-mendations, either to individual observers or to the association itself.

Although we have been discussing times of minimum for eclipsing binaries, there is also some interest in times of maximum light for Cepheid and RR Lyrae variables, especially the latter, which undergo period changes (Tsesevich 1969) which are just as poorly understood as they are for binary stars.

## 16.4 Discovering New Variables

Discovering new variable stars probably will continue to be scientifically useful to some extent for many years to come. There are already an extremely large number of variables known. The *Third Edition of the General Catalogue of Variable Stars* altogether lists 20,437 stars designated as variable as of 1968. The first three supplements increase this total by 5,405 making a grand total of 25,842. The total by now is even larger. With the advent of photoelectric photometry more and more variables are being discovered with small amplitudes, some around $0^m01$. Even Commission 27 (Variable Stars) of the I.A.U. has not yet come to grips with the fact that, in principle, all stars must be variable, the only meaningful distinction being one of degree. For example, no one doubts now that even our Sun is variable, at least by $0^m001$ and perhaps by more than $0^m01$. In general, bright stars will be more useful to astronomy than fainter stars simply because more detailed information can be obtained by observation if more light is available for analysis. Therefore, searching for variables among the bright stars, i.e., those 9110 in the *Yale Bright Star Catalogue*, would be a meaningful observing program for amateurs and small college observatories with telescopes of small aperture capable of accurate differential photoelectric photometry. Among the bright stars, however, one should expect not many variables of large amplitude to have gone undiscovered this long. On the other hand, in the amplitude range $0^m03 < \Delta m < 0^m3.$, large enough for photoelectric photometry but not large enough for photographic or visual photometry, there could well be hundreds of such variables.

Recently, in the process of preparing the fourth edition of the *Yale Bright Star Catalogue*, Hoffleit (1979) has drawn attention to the fact that 1261 of the 9110 have been suspected of variability at one time or another. Many of these suspects will prove to be red herrings but not all of them. One indication of this is the excellent but still unpublished Ph.D. thesis of Winzer (1974). Here he reported on his search for variability among all of the Ap stars in the *Yale Bright Star Catalogue* having spectral types in the range B8 to A2. Out of the 99 he observed, 15 ( = 15%) were already known variables and 44 ( = 44%) were discovered by him to be variable. Their amplitudes ranged between $0^m01$ and $0^m10$.

## 16.5 Tumbling Asteroids

Because most asteroids are considerably non-spherical, most vary in brightness (on a time scale of hours) as they rotate. Suitable analysis of photometry can yield not only the rotational period but also the orientation of the rotational axis. The photometry itself is significantly more difficult than that of variable stars. Because asteroids move with respect to the background stars, it is much more difficult to prepare suitable finding charts and, moreover, one must use different comparison stars at different times. Analysis of resulting light curves is also more difficult because one must correct the times for the ever-changing light travel time and because one must correct the apparent magnitudes for the ever-changing sun-asteroid and earth-asteroid distances. Nevertheless, such work is of considerable value and, judging by the number of entries in Table 16–1, of considerable interest. There is a publication called the *Minor Planet Bulletin* (Binzel 1981, Genet 1983a) devoted primarily to this sort of investigation.

## 16.6    Occultations

Our book has said very little about photoelectric photometry of lunar, planetary, or asteroid occultation events, emphasizing variable stars instead. Nevertheless, there has been and still is activity in the field of photoelectric timing of occultation events. Good references in this area are by Evans et al. (1979) for lunar occultations and Elliott (1979) for planetary occultations. Several chapters in the *Solar System Photometry Handbook* edited by Genet (1983c) are relevant. And there is an organization called I.O.T.A. which works specifically to coordinate the precise timing of these occultations of stars by asteroids.

Accomplishments resulting from occultation studies have included the following: resolving previously unresolved double stars, measuring stellar diameters, measuring the diameters of asteroids, determining the diameter of Pluto, determining the profile of the moon's limb, and discovering the rings of Uranus and, most recently, possible rings around Neptune.

## 16.7    Photoelectric Sequences

By emphasizing differential photometry of variable stars, our book has said nothing about establishing photoelectric sequences. This necessary and worthwhile activity, in which the photoelectric photometrist comes to the aid of the visual observer, involves determining accurate magnitudes of those stars which visual observers use as a sequence of comparison stars when estimating brightness differences by eye. Anyone interested in doing work of this sort should contact headquarters of the relevant organization, such as the A.A.V.S.O. This activity is not trivial or easy. In addition to determining magnitudes photoelectrically on some standard photometric system, one must deal with the problem of relating those standard photoelectric magnitudes to actual "visual magnitudes," i.e., magnitudes which will correspond to what the dark-adapted human eye sees. An excellent review of this problem is given by Howarth (1979).

## 16.8    The RS Canum Venaticorum Binaries

We would like to outline one specific observing program in detail: systematic differential photoelectric photometry of variable RS CVn binaries. This was presented in an early issue of the *I.A.P.P.P. Communications* (Hall 1980). Although at this present moment it still represents a scientifically worthwhile problem and probably will continue to be so for some time, we prefer to view it as an instructive example of what a feasible and appropriate observing program might be.

Another motivation for selecting this particular one is that the RS CVn-type, along with the closely related BY Draconis-type variables, represent one of the few genuinely new type of variable stars discovered in the last 20 years. Starting with the *First Supplement to the Third Edition*, the *General Catalogue of Variable Stars* now identifies variables of this type with the designation BY.

There are now several references describing these fascinating stars: two review papers (Hall 1976, 1981), two articles in popular astronomy magazines (Zeilik et al. 1979, Hall 1983a), and a list of still unsolved problems (Hall, 1987). Recently, there was a N.A.T.O. Advanced Study Institute which discussed the exciting new topic of the occurrence of solar-type phenomena (sunspots, chromospheres, magnetic fields, spot cycles, flares, coronas, X–ray emission, radio outbursts, and solar wind) on stars other than the sun. The RS CVn binaries figured prominently in the published proceedings of the Institute (Bonnet and Dupree, 1981) as well as in the first five of the so-called Cambridge Cool Star Workshops.

The RS Canum Venaticorum-type binaries are composed of two stars fairly close together but not enough so that either one overflows its Roche lobe, i.e., they are detached binaries. Their orbital periods range from around a day or two to around a month or two, depending on how the group is formally defined. Both stars are cool enough to have convection in their outer layers, as the sun does. In a typical system both have begun to evolve off the main sequence, one more so than the other, with the former now being a subgiant. The subgiant, having been forced to rotate synchronously with the orbital motion, is now rotating many times faster than the sun and much faster than a single subgiant of the same size would be. For some reason not yet understood too well, this rapid rotation in a star with a deep convective layer gives rise to solar-type phenomena on its surface far more extreme than what our sun ever experiences. Its chromospheric emission lines indicate a chromosphere far more active than the sun's. Detailed spectra taken in the far ultraviolet region of the spectrum by the Copernicus Satellite and by the International Ultraviolet Explorer Satellite indicate giant loop prominences containing a thousand times more gas than the largest solar prominences ever do. Whereas never more than 1% of the sun's surface is covered by sunspots, even during maxima of the 11–year cycle, it has been deduced that the subgiants in most of the RS CVn-type binaries often have up to 25% or 30% of one hemisphere covered with dark spots. Their coronas are denser and much hotter than the sun's corona is, around 10,000,000°K rather than around 1,000,000°K. Consequently, they emit X–ray radiation roughly 10,000 times more intense than the sun's. Because of this, the RS CVn-type binaries are proving to account for roughly 10% percent of all the known X–ray sources in the sky—stellar, galactic, and extragalactic. Their coronas are evaporating, giving rise to a rapidly out-flowing plasma of hot ionized gas similar to the solar wind but about 10,000 times stronger. Many of the RS CVn binaries are known sources of radio emission also. This emission is highly variable and flare-like in character, similar in many ways to the radio emission from the sun, except that outbursts on RS CVn binaries can be 1,000,000 or 100,000,000 times more intense intrinsically than the strongest ones ever recorded from the sun. Probably as a consequence of the large rate of mass outflow from the subgiant and its interaction with the corona and magnetic field of the other star, the orbital period itself in many of these binaries is varying sometimes enough to make an ephemeris determined only ten years before predict times of eclipse which are in error by an hour or more.

The reason that the RS CVn binaries tend to be variable stars is the curious fact that the dark spots are not distributed uniformly or randomly across the subgiant's surface, but rather are concentrated in one (or maybe two) large groups. Therefore, as the binary revolves and the synchronously rotating subgiant turns with it, we see alternately the spotted and then the unspotted hemisphere. This produces light variations which can be as large as $0^m45$ in the extreme case of II Peg (Byrne 1986), is around $0^m1$ for many systems, and for only very few systems is so small that none has ever been detected. This type of light variation which occurs in RS CVn binaries is usually referred to as the wave. The period of the light variation for most of the RS CVn binaries is very close to the orbital period but never exactly the same. The small difference, generally less than 1%, causes the wave to drift slightly or migrate in a light curve plotted versus orbital phase. Typically, the wave will complete one migration cycle, i.e., wave minimum will come back to the same value of orbital phase in something like 10 years, in which case 10 years would be referred to as the migration period. The exact shape of the wave sometimes changes from cycle to cycle and often changes from year to year, presumably as the shape and/or location and/or size of the spot group changes. For a few RS CVn binaries which have been observed faithfully for many years, the wave amplitude has changed gradually year by year in such a way as to suggest that a spot cycle, perhaps analogous to our sun's 11–year cycle, is operating in these stars also. If an RS CVn binary is also an eclipsing binary, and quite many are, then the wave is interrupted whenever an eclipse occurs. One of the most puzzling unanswered questions about the RS CVn binaries is why the spots tend to occur in groups and why those groups seem to be so long-lived. Spot groups in some systems have been shown to have persisted and maintained their identities for at least a decade and probably longer. It is ironic, therefore, that if the spots were distributed uniformly or randomly, then the RS CVn binaries would not vary in light as the binary revolved and very likely we would be altogether unaware of the existence of the spots even now.

The table in Hall (1980) listed two dozen bright RS CVn binaries known to be or suspected of being variable with a reasonably large amplitude. Also listed were a suggested comparison star for each and remarks about any nearby visual companions which should be either included in or excluded from the diaphragm deliberately during photometry. The orbital period of each was given, when known, since in most cases the light varies with approximately that same period. The V magnitudes given, which refer to the approximate average light level, showed that half are of naked eye brightness.

The observing program was very simple. Each clear night on which you can observe, you should go through the basic sequence, thus getting three differential magnitudes between the variable and its comparison. Do so for as many variables on the list as you can, provided they are in the sky for that season. Only for binaries with periods shorter than a couple of days would it be worthwhile to go through the basic sequence twice in any one night. Observations in V only are sufficient for this program. Pretty much as described, V bandpass differential photometry of all two dozen of these RS CVn binaries was obtained by more than a dozen small observatories for three or four years and sent to the program coordinator for analysis and publication. Looking through the "Papers Published by Amateurs" feature in the *I.A.P.P.P. Communications*, we see more than 30 papers pertaining to those 24 stars. The number of different amateurs included as co-authors is almost three dozen.

Let us point out how, vis-à-vis the factors which should be considered when selecting an observing program (described in Chapter 15), the RS CVn program described above was in very many respects ideal. It was very well-suited for both the amateur and small college observatory with a telescope of small or modest aperture situated where it is readily accessible but hampered by mediocre sky conditions, such as bright sky background and relatively few cloudless nights each year.

1. Many are very bright and thus suitable for small telescopes or observatories with bright sky background. Of the two dozen listed by Hall (1980) almost half are of naked eye brightness. Of the 70 listed by Hall (1981), 25 are included in the *Yale Bright Star Catalogue*, and thus are also of naked eye brightness.

2. Of those same two dozen, all have amplitudes over $0^{\rm m}1$. Thus, their light curves can be defined adequately by differential photoelectric photometry which is accurate at about the $\pm 0^{\rm m}01$ level.

3. It is to some extent an additional advantage that their amplitudes are not larger than about $0^{\rm m}3$. Variables with amplitudes greater than about $0^{\rm m}4$ or $0^{\rm m}5$ have already been discovered in large numbers by photographic variable star survey techniques and, once discovered, often followed up promptly by photoelectric photometry.

4. Because the light variations are slow and gradual, one observation per night is sufficient and small gaps in the light curve can be tolerated. Such would not be true of an eclipsing binary during eclipse.

5. They are periodic variables and their light curves repeat fairly well from one cycle to the next. Therefore, because it is meaningful to combine several cycles onto the same light curve, it is not essential to observe all parts of each individual cycle.

6. On the other hand, slight changes from cycle to cycle and more significant changes from year to year do occur and are of scientific interest in learning how spot groups change size, shape, and location (Bopp and Noah 1980, Dorren et al. 1981). A periodic variable whose light curve does not change at all can be observed well once and for all, and then forgotten. Because the RS CVn light curves are always changing, we will not be finished with them until we understand those changes completely, and that day is very far into the future.

7. They are well suited for convenient single-filter photometry. This is because the light curve shapes and amplitudes are very similar in all bandpasses of the optical window. This virtual constancy of the color index also makes correction for color-dependent extinction and for transformation quite simple.

8. Theoreticians want photometry of these variables very badly. Just now theoretical astrophysicists are becoming keenly interested in this remarkable spot activity: its physical nature, its origin, its large scale, its various periodicities, etc. A good reference is Shore and Hall (1980). But serious systematic photometry of these stars has begun only relatively recently and, with some of the interesting

periodicities (like the migration period and the suspected spot cycles) being quite long, the photometric data they need are not available to them yet.

9. The long orbital periods and the even longer time scales of the possible spot cycles make these variables almost impossible for professionals to handle during the relatively brief scheduled observing runs at large observatories. However, these same conditions make these variables perfectly suited for the astronomer with ready access to his backyard observatory all year long.

10. Because of the activity in the radio, far-ultraviolet, and X-ray spectral regions, N.A.S.A. and the large radio observatories have become very interested in the RS CVn binaries, often needing solid night-by-night simultaneous ground-based photometric backup during a scheduled pointing or an unusual radio outburst. Because those same target stars are bright enough for photometry with 8-inch telescopes, the coordinated collective effort of amateur and small college observatories all across the continent can defeat cloudcover and provide that backup which smaller numbers of large professional observatories cannot.

# REFERENCES

Abt, H. A. 1980, P.A.S.P. **92**, 249.

Africano, J. and Quigley, R. 1977, J.A.A.V.S.O. **6**, 53.

Albrecht, R., Boyce, P., and Chastain, J. 1971, P.A.S.P. **83**, 683.

Allen, C. W. 1973, *Astrophysical Quantities* (London: Athlone Press).

Argue, A. N. and Butler, H. E. 1952, the Observatory **72**, 31.

Asknes, K. and Franklin, F. 1985, I.A.P.P.P. Comm. No. 19, 23.

Batten, A. H. 1973, *Binary and Multiple Systems of Stars* (New York: Pergamon).

Baum, W. A. 1962, in *Astronomical Techniques*, edited by W. A. Hiltner (Chicago: University of Chicago Press).

Bensammar, S. 1978, A. & A. **65**, 199.

Bernacca, P. L., Canton, G., Stagnu, R., Lesciutta, S., and Sedmak, G. 1978, A. & A. **70**, 821.

Bemporad, A. 1904, Heidelberg Mitteilungen No. 4.

Bessell, M. S. 1979, P.A.S.P. **91**, 589.

Binzel, R. P. 1981, I.A.P.P.P. Comm. No. 5. 19.

Bisard, W. and Osborn, W. 1984, I.A.P.P.P. Comm. No. 17, 73.

Bonnet, R. M. and Dupree, A. K. 1981, *Solar Phenomena in Stars and Stellar Systems* (Dordrecht: Reidel).

Bopp, B. W. and Noah, P. V. 1980, P.A.S.P. **92**, 717.

Borucki, W. J. and Young, A. 1984, N.A.S.A. Conf. Publ. No. 2350.

Boyd, L. J. 1984, I.A.P.P.P. Comm. No. 17, 1.

Boyd, L. J., Genet, R. M. and Hall, D. S. 1984, I.A.P.P.P. Comm. No. 15, 20.

Brettman, O. H., Fried, R. E., DuVall, W. M., Hall, D. S., Poe, C. H., and Shaw, J. S. 1983, I.B.V.S. No. 2389.

Burnet, M. and Rufener, F. 1979, A. & A. **74**, 54.

Byrne, P. B. 1986, I.B.V.S. No. 2951.

Campbell, L. 1955, *Studies of Long Period Variables* (Cambridge: A.A.V.S.O.).

Campbell, L. and Jacchia, L. 1941, *The Story of Variable Stars* (Philadelphia: the Blakisten Company).

Chambliss, C. R., Hall, D. S., Landis, H. J., Louth, H., Olson, E. C., Renner, T. R., and Skillman, D. R. 1978, A.J. **88**, 1514.

Clark, D. H. and Stephenson, F. R. 1977, *The Historical Supernovae* (New York: Pergamon Press).

Clarke, D. 1978, Nature **274**, 670.

Crawford, D. L. 1969, in *Stellar Astronomy*, Volume I, edited by H. Y. Chiu, R. L. Warasila, J. L. Remo (New York: Gordon and Breach), chapter I-3.

Crawford, D. L. and Barnes, J. V. 1970, A.J. **75**, 978.

Crawford, D. L., Golson, J. C., and Landolt, A. U. 1971, P.A.S.P. **83**, 652.

236

Crawford, D. L. and Mander, J. 1966, A.J. **71**, 114.

Cummings, E. E. 1921, P.A.S.P. **33**, 214.

Davidson, J. K., Neff, J. S., and Enemark, D. C. 1976, P.A.S.P. **88**, 209.

DeBiase, G. A., Lucio, P., Pucillo, M., and Sedmak, G. 1978, Applied Optics **17**, 435.

DeBiase, G. A. and Sedmak, G. 1974, A. & A. **33**, 1.

Dorren, J. D., Siah, M. J., Guinan, E. F., and McCook, G. P. 1981, A.J. **86**, 572.

DuPuy, D. L. 1981, P.A.S.P. **93**, 144.

DuPuy, D. L. 1983a, I.A.P.P.P. Comm. No. 11, 23.

DuPuy, D. L. 1983b, P.A.S.P. **95**, 86.

Eaton, J. A. 1983, I.A.P.P.P. Comm. No. 14, 60.

Eggleton, P. P. and Pringle, J. E. 1985, *Interacting Binaries* (Dordrecht: Reidel).

Elliot, J. L. 1979, A.R.A.A. **17**, 445.

Evans, D. S., Barnes, T. G., and Lacy, C. H. 1979, Sky and Telescope **58**, 130.

Fernie, J. D. 1975, Operating Manual for the D.D.O. Chopping Photometer.

Fernie, J. D. 1976, P.A.S.P. **88**, 969.

Fernie, J. D. 1979, David Dunlap Doings **12**, 7.

Fernie, J. D. 1983a, I.A.P.P.P. No. 13, 16.

Fernie, J. D. 1983b, P.A.S.P. **95**, 782.

Fossat, E., Harvey, J., Hausman, M. and Slaughter, C. 1977, A. & A. **59**, 279.

Genet, R. M. 1982, *Real Time Control with Microcomputers* (Indianapolis: H. W. Sams).

Genet, R. M. 1983a, I.A.P.P.P. Comm. No. 12, 74.

Genet, R. M. 1983b, *Microcomputers in Astronomy*, Volume I (Fairborn: Fairborn Press).

Genet, R. M. 1983c, *Solar System Photometry Handbook* (Richmond: Willmann-Bell).

Genet, R. M. 1986, I.A.P.P.P. Comm. No. 25, chapter 1.

Genet, R. M. and Genet, K. A. 1984, *Microcomputers in Astronomy*, Volume II (Fairborn: Fairborn Press).

Geyer, E. H. and Hoffman, M. 1975, A. & A. **38**, 359.

Ghedini, S. 1982, *Software for Photometric Astronomy* (Richmond: Willmann-Bell).

Glasby, J. S. 1969, *Variable Stars* (Cambridge: Harvard University Press).

Glasby, J. S. 1970, *The Dwarf Novae* (New York: Elsevier).

Glasby, J. S. 1974, *The Nebular Variables* (New York: Pergamon Press).

Golay, M. 1974, *Introduction to Astronomical Photometry* (Boston: Reidel).

Grec, G. and Fossat, E. 1979, A. & A. **77**, 351.

Gurzadyan, G. A. 1980, *Flare Stars* (Oxford: Pergamon Press).

Hall, D. S. 1976, I.A.U. Colloquium No. 29, Part I, 287.

Hall, D. S. 1980, I.A.P.P.P. Comm. No. 3, 1.

Hall, D. S. 1981, in *Solar Phenomena in Stars and Stellar Systems*, edited by R. M. Bonnet and A. K. Dupree (Dordrecht: Reidel).

Hall, D. S. 1983a, Astronomy **11**, No. 2, 66.

Hall, D. S. 1983b, I.A.P.P.P. Comm. No. 11, 3.

Hall, D. S. 1983c, I.A.P.P.P. Comm. No. 12, 74.

Hall, D. S. 1983d, I.A.P.P.P. Comm. No. 14, 14.

Hall, D. S. 1987, Publ. Astr. Inst. Czechoslovakia **70**, 77.

Hall, D. S. and Genet, R. M. 1982, *Photoelectric Photometry of Variable Stars* (Fairborn: I.A.P.P.P.).

Hall, D. S., Genet, R. M., and Thurston, B. L. 1986, I.A.P.P.P. Comm. No. 25.

Hall, J. S. 1932, Proceedings of the National Academy of Science **18**, 365.

Hardie, R. H. 1959, Ap.J. **130**, 663.

Hardie, R. H. 1962, in *Astronomical Techniques*, edited by W. A. Hiltner (Chicago: University of Chicago Press), p. 178.

Hayes, D. S. and Latham, D. W. 1975, Ap.J. **197**, 593.

Hearnshaw, J. B. and Cottrell, P. L. 1986, I.A.U. Symposium No. 118.

Henden, A. A. and Kaitchuck, R. H. 1982, *Astronomical Photometry* (New York: Van Nostrand Reinhold).

Hoffleit, D. 1979, J.A.A.V.S.O. **8**, 34.

Hoffmeister, C. 1970, *Veränderliche Sterne* (Leipzig: Verlag Barth).

Hoffmeister, C., Richter, G., and Wenzel, W. 1985, *Variable Stars* (Berlin: Springer Verlag).

Honeycutt, R. K., Kephart, J. E., and Henden, A. A. 1978, Sky and Telescope **56**, 495.

Howarth, I. D. 1979, J.A.A.V.S.O. **8**, 26.

Hudson, K. I., Chiu, H. Y., Maran P., Stuart, F. E., Vokac, P. R. 1971, Ap.J. **165**, 573.

Iriarte, B., Johnson, H. L., Mitchell, R. I., and Wisniewski, W. K. 1965, Sky and Telescope **30**, 21.

Johnson, H. L. 1952, Ap.J. **116**, 640.

Johnson, H. L. 1954, Ap.J. **119**, 181.

Johnson, H. L. 1963, in *Basic Astronomical Data*, edited by K. A. Strand (Chicago: University of Chicago Press), p. 204.

Johnson, H. L. and Harris, D. L. 1954, Ap.J. **120**, 196.

Johnson, H. L. and Kunckles, C. F. 1955, Ap.J. **122**, 209.

Johnson, H. L., Mitchell, R. I., Iriarte, B., and Wisniewski, W. Z. 1966, Communications of the Lunar and Planetary Laboratory No. 63.

Johnson, H. L. and Morgan, W. W. 1951, Ap.J. **114**, 522.

Johnson, H. L. and Morgan, W. W. 1953, Ap.J. **117**, 313.

Jones, E. L. 1980, Sky and Telescope **60**, 333.

Kaitting, M. K. 1984, I.A.P.P.P. Comm. No. 17, 64.

Kibrick, R., Rickets, T., and Robinson, L. 1979, Proceedings of the S.P.O.I.E. **172**, 403.

Kopal, Z. 1978, *Dynamics of Close Binary Systems* (Dordrecht: Reidel).

Kopal, Z. 1979, *Language of the Stars* (Dordrecht: Reidel).

Kron, G. E. 1947, P.A.S.P. **59**, 173.

Kron, G. E. 1974, in *Methods of Experimental Physics*, Volume **12**, Part A, edited by N. Carleton (New York: Academic Press), Chapter 6.

Kron, G. E. 1981, I.A.P.P.P. Comm. No. 5, 4.

Kron, G. E. and Fellgett, P. B. 1955, P.A.S.P. **67**, 334.

Kron, G. E. and Smith, J. L. 1951, Ap.J. **113**, 324.

Kron, G. E., White, H. S., and Gascoigne, S. C. P. 1953, Ap.J. **118**, 502.

Kurarkin, B. V. 1970, *Pulsating Stars* (New York: Wiley).

238

Lallemand, A. 1936, Compte Rendu Academy Science Paris **203**, 243.

Landis, H. J., Louth, H., and Hall, D. S. 1985, I.B.V.S. No. 2662.

Landolt, A. U. and Blondeau, K. L. 1972, P.A.S.P. **84**, 784.

Larsson-Leander, G. 1969, in *Non-Periodic Phenomena in Variable Stars*, edited by L. Detre (Budapest: Academic Press), p. 443.

Larsson-Leander, G. 1970, Arkiv for Astronomi **5**, 253.

Lasker, B. M. 1972, P.A.S.P. **84**, 207.

Lines, R. D., Lines, H. C., Boyd, L. J., and Genet, R. M. 1985, I.A.P.P.P. Comm. No. 21, 46.

Lines, R. D. and Hall, D. S. 1981, I.B.V.S. No. 2013.

McNall, J. F., Miedaner, T. L., and Code, A. D. 1968, A.J. **73**, 756.

Mallama, A. D. 1974a, J.A.A.V.S.O. **3**, 11.

Mallama, A. D. 1974b, J.A.A.V.S.O. **3**, 49.

Maran, S. P. 1969, in *Stellar Astronomy*, Volume I, edited by H. Y. Chiu, R. L. Warasila, J. L. Remo (New York: Gordon and Breach), p. 323.

Melsheimer, D. F. and Genet, R. M. 1984, I.A.P.P.P. Comm. No. 15, 33.

Merrill, P. W. 1938, *The Nature of Variable Stars* (New York: Macmillan).

Minchin, G. M. 1895, Proceedings of the Royal Society **58**, 142.

Minchin, G. M. 1896, Proceedings of the Royal Society **59**, 231.

Moore, C. H. and Rather, E. D. 1973, Proceedings of the I.E.E.E. **61**, 1346.

Nather, R. E. 1973, Vistas in Astronomy **15**, 91.

Nelson, B. and Young, A. 1970, P.A.S.P. **82**, 699.

Oliver, J. P. 1982a, in *Photoelectric Photometry of Variable Stars*, edited by D. S. Hall and R. M. Genet (Fairborn: I.A.P.P.P.), p. 5-16.

Oliver, J. P. 1982b, in *Photoelectric Photometry of Variable Stars*, edited by D. S. Hall and R. M. Genet (Fairborn: I.A.P.P.P.), figure 5-7.

Olson, E. C. 1984, I.A.P.P.P. No. 15, 5.

Payne-Gaposchkin, C. 1964, *The Galactic Novae* (New York: Dover).

Pecker, J. C. 1970, *Space Observatories* (Dordrecht: Reidel).

Percy, J. R. 1986, *The Study of Variable Stars Using Small Telescopes* (Cambridge: Cambridge University Press).

Petit, M. 1982, *Les Etoiles Variables* (Paris: Masson).

Reddish, V. C. 1966, Sky and Telescope **32**, 124.

Robinson, L. B. 1975, A.R.A.A. **13**, 165.

Sahade, J. and Wood, F. B. 1978, *Interacting Binary Stars* (New York: Pergamon Press).

Sanders, W. H. and Persha, G. 1983, I.A.P.P.P. Comm. No. 14, 19.

Seeds, M. A. 1970, Ph.D. Thesis, Indiana University.

Shao, C. Y. and Young, A. T. 1965, A.J. **70**, 726.

Shore, S. N. and Hall, D. S. 1980, I.A.U. Symposium No. 88, 389.

Skillman, D. R. 1981, Sky and Telescope **61**, 71.

Skillman, D. R. 1982, J.A.A.V.S.O. **11**, 57.

Smith, D. W. 1985, I.A.P.P.P. Comm. No. 21, 3.

Smith, S. 1932, Contributions from Mount Wilson Observatory No. 457.

Snell, C. M. and Heiser, A. M. 1968, P.A.S.P. **80**, 336.

Soderhjelm, S. 1980, A. & A. **89**, 100.

Sorvari, J. M. 1975, P.A.S.P. **87**, 443.

Stebbins, J. 1940, P.A.S.P. **52**, 235.

Stebbins, J. and Whitford, A. E. 1943, Ap.J. **98**, 20.

Strohmeier, W. 1972, *Variable Stars* (New York: Pergamon Press).

Strömgren, B. 1966, A.R.A.A. **4**, 433.

Titus, J. A. 1979, *TRS-80 Interfacing* (Indianapolis: H. W. Sams).

Trueblood, M. and Genet, R. M. 1985, *Microcomputer Control of Telescopes* (Richmond: Willmann-Bell).

Tsesevich, V. P. 1969, *RR Lyrae Stars* (New York: Wiley).

Tsesevich, V. P. 1970, *Variable Stars and Methods for their Investigation* (Moscow: Pedagogika).

Walter, K. 1985, I.A.P.P.P. Comm. No. 22, 30.

Weiler, E. J., Owen, F. N., Bopp, B. W., Schmitz, M., Hall, D. S., Fraquelli, D. A., Piirola, V., Ryle, M., and Gibson, D. M. 1978, Ap.J. **225**, 919.

Whitford, A. E. 1938, in *Procedures in Experimental Physics* (London: Prentice-Hall).

Whitford, A. E. 1962, in *Handbuch der Physik*, edited by S. Flugge (Berlin: Springer Verlag).

Wing, R. F. and Hall, D. S. 1983, I.A.P.P.P. Comm. No. 14, 57.

Winzer, J. 1974, Ph.D. Thesis, University of Toronto.

Wolpert, R. C. and Genet, R. M. 1983, *Advances in Photoelectric Photometry*, Volume I (Fairborn: Fairborn Press).

Wolpert, R. C. and Genet, R. M. 1984, *Advances in Photoelectric Photometry*, Volume II (Fairborn: Fairborn Press).

Wood, F. B. 1953, *Astronomical Photoelectric Photometry* (Washington, D.C.: A.A.A.S.).

Wood, F. B. 1963, *Photoelectric Astronomy for Amateurs* (New York: Macmillan).

Young, A. T. 1974, in *Methods of Experimental Physics*, Volume **12**, Part A, edited by N. Carleton (New York: Academic Press), Chapters 1, 2, and 3.

Zeilik, M., Hall, D. S., Feldman, P. A., Walter, F. 1979, Sky and Telescope **57**, 132.

# Appendix A

# Abbreviations and Acronyms

**A & A** = *Astronomy and Astrophysics*

**AAAS** = American Association for the Advancement of Science

**AAVSO** = American Association of Variable Star Observers

**AC** = alternating current

**AD** = Analog Devices

**AFOEV** = l'Association Francaise d'Observateurs d'Etoiles Variables

**AJ** = *Astronomical Journal*

**Amp/Disc** = amplifier/discriminator

**ApJ** = *Astrophysical Journal*

**APT** = automatic photoelectric telescope

**ARAA** = *Annual Reviews of Astronomy and Astrophysics*

**ASS** = *Astrophysics and Space Science*

**BASIC** = Beginner's All-Purpose Symbolic Instruction Code

**BAV** = Berliner Arbeitsgemeinschaft für Veránderliche Sterne

**BBSAG** = Bedeckungsveränlichen Beobachter der Schweizerischen Astronomischen Gesellschaft

**BCD** = binary-coded decimal

**BNC** = baby "N" connector

**BUD** = Bud [Haas]

**CB** = Corning blue

**CCD** = charge-coupled device

**CMOS** = complimentary metal oxide semiconductor

**CPS** = counts per second

**CR** = Corning red

**CTIO** = Cerro Tololo InterAmerican Observatory

**DAO** = Dominion Astrophysical Observatory

**DATUG** = Dayton Area TRS-80 User's Group

**DC** = direct current

**DDO** = David Dunlap Observatory

**DEC** = declination

**DFM** = D. F. Melsheimer

**DIP** = dual in-line package

**DPV** = digital panel voltmeter

**ECL** = emitter-coupled logic

**EMF** = electro-motive force

**EMI** = Electronic Music Industry

**ERMA** =        extended        red
   multi-alkalai

**FET** = field effect transistor

**FORTH** = Fourth (Genereation)

**FORTRAN** = Formula Translation

**FS** = full scale

**FWHM** = full width at half maximum

**GND** = ground

**GRD** = ground

**HD** = Henry Draper [Catalogue]

**HP** = Hewlett-Packard

**HPO** = Hopkins-Phoenix Observatory

**HR** = Harvard Revised [Photometry Catalogue]

**HVPS** = high voltage power supply

**IAPPP** = International Amateur-Professional Photoelectric Photometry

**IAU** = International Astronomical Union

**IBM** = International Business Machines

**IBVS** = *Information  Bulletin  on Variable Stars*

**IC** = integrated circuit

**ICBM** = Intercontinental Ballistic Missile

**IEEE** = Institute of Electrical and Electronic Engineers

**I/O** = input/output

**IOTA** = International Occultation Timing Association

**IR** = infrared

**ITT** = International Telephone and Telegraph

**JAAVSO** = *Journal of the American  Association  of  Variable Star Observers*

**JD** = Julian date

**KIM** = Keyboard Interface Module

**KPNO** = Kitt Peak National Observatory

**LED** = light-emitting diode

**LSI** = large-scale integration

**LST** = local sidereal time

**MHV** =   miniature   high   voltage
   [connector]

**MJD** = modified Julian date

**MMT** = multiple-mirror telescope

**NASA** = National Aeronautics and Space Administration

**NATO** = North Atlantic Treaty Association

**NEA** = Nouveau Era Astronomique

**NEP** = noise-equivalent power

**NIM** = National Industrial Manufacturers

**NPN** = Negative-Positive-Negative

**OAO** = Orbitting Astronomical Observatory

**OK** = okey

**OpAmp** = operational amplifier

**PASP** = *Publications of the Astronomical Society of the Pacific*

**PC** = personal computer

**PC** = printed circuit

**PCB** = printed circuit board

**PCV** = poly-chlorinated vinyl

**PDS** = photographic density scanner

**PE** = photoelectric

**PEP** = photoelectric photometry

**PG** = photographic

**PIN** = positive-intrinsic-negative

**PIT** = programmable interval timer

**PIV** = peak inverse voltage

**PMT** = photomultiplier tube

**PN** = positive-negative

**POT** = potentiometer

**PP** = peak-to-peak

**PPI** = programmable peripheral interface

**PPM** = parts per million

**Preamp** = preamplifier

**PVH** = Pleione Valtozocsillag-Eszlelo Halozat

**QE** = quantum efficiency

**RA** = right ascension

**RASNZ** = Royal Astronomical Society of New Zealand

**RCA** = Radio Corporation of America

**RF** = radio frequency

**RFI** = radio frequency interference

**RI** = red, infrared

**RMS** = root mean square

**RTV** = room temperature vulcanizer

**SAO** = Smithsonian Astrophysical Observatory

**SHA** = sample-and-hold amplifier

**S/N** = signal-to-noise ration

**S & T** = *Sky and Telescope*

**SPOIE** = Society of Photo-Optical Instrumentation Engineers

**SSP** = solid-state photometer

**SW** = short wave

**TE** = thermoelectric

**TNC** = threaded "N" connector

**TRS** = Tandy Radio Shack

**TTL** = transistor-transistor logic

**UBV** = ultraviolet, blue, visual

**UHF** = ultrahigh frequency

**USAFA** = United States Air Force Academy

**UT** = Universal Time

**UV** = ultraviolet

**uvby** = ultraviolet, violet, blue, yellow

**V/F** = voltage-to-frequency

**VHF** = very high frequency

**WWV** = WWV

**ZT** = zone time

# Appendix B

# Addresses

American Association of Variable
Star Observers
25 Birch Street
Cambridge, Massachusetts 02138

American Astronomical Society
2000 Florida Avenue, N.W.
Suite 300
Washington, D.C. 20009

Analog Devices, Inc.
One Technology Way
P.O. Box 9106
Norwood, Massachusetts 02062

Automatic Photoelectric Telescope
Service
Mr. Louis J. Boyd
629 North 30th Street
Phoenix, Arizona 85008

Barr-Brown Research Corporation
6730 South Tucson Boulevard
Tucson, Arizona 85734

B.A.V.
Mr. Rainer Lukas
Sternwarte
Munsterdamm 90
D-1000 Berlin 41
West Germany

B.B.S.A.G.
Kurt Locher
Rebrain 39
CH-8624 Grut
Switzerland

BUD Industries, Inc.
4605 East 355th Street
P.O. Box 431
Willoughby, Ohio 44094

Celestron International
2835 Columbia Street
Box 3578
Torrance, California 90503

Corning Glass Works
Houghton Park
Corning, New York 14830

David Dunlap Observatory
Richmond Hill, Ontario
Canada L4C 4Y6

D.F.M. Engineering
1812 Valtec Lane
Bay 3
Boulder, Colorado 80301

Edmund Scientific
101 East Gloucester Pike
Barrington, New Jersey 08007

Hamamatsu Corporation
420 South Avenue
Middlesex, New Jersey 08846

Heathkit
Heath Company
Benton Harbor, Michigan 49022

Hewlett-Packard Company
5201 Tollview Drive
Rolling Meadows, Illinois 60008

Hopkins-Phoenix Observatory
Mr. Jeffrey L. Hopkins
7812 West Clayton Drive
Phoenix, Arizona 85033

Hurst Manufacturing Corporation
Box 326
Princeton, Indiana 47670

I.B.V.S.
Prof. Bela Szeidl, Editor
Konkoly Observatory
H-1525 Budapest XII. Box 67.
Hungary

I.B.M.
400 Columbia Avenue
Valhalla, New York 10595

I.A.P.P.P.
Dyer Observatory
Vanderbilt University
Nashville, Tennessee 37235

I.A.U.
Dr. Jean-Paul Swings
61 avenue de l'Observatoire
F-75014 Paris
France

I.O.T.A.
Dr. David W. Dunham
P.O. Box 7488
Silver Springs, Maryland 20907

I.T.T.
Electro Optical Products Division
3700 East Pontiac Street
Fort Wayne, Indiana 46803

Intersil, Inc.
10710 Tantau Avenue
Cupertino, California 95014

Jameco Electronics
1021 Howard Avenue
San Carlos, California 94070

Keithley Instruments, Inc.
28775 Aurora Road
Cleveland, Ohio 44139

Kitt Peak National Observatory
P.O. Box 26732
Tucson, Arizona 85726

LeCroy Corporation
700 Chestnut Ridge Road
Chestnut Ridge, New York 10977

Thomas Mathis Company
830 Williams Street
San Leandro, California 94577

Meade Instruments Corporation
1675 Toronto Way
Costa Mesa, California 92626

Minor Planet Bulletin
Mr. Derald D. Nye
Route 7, Box 15
Tucson, Arizona 85747

Optec, Inc.
199 Smith Street
Lowell, Michigan 49331

Pacific Precision Instruments
1040 Shary Court
Concord, California 94518

R.C.A.
Electro Optics and Devices
Lancaster, Pennsylvania 17604

Royal Astronomical Society of New
Zealand
Variable Star Section
P.O. Box 3093
Greerton, Tauranga
New Zealand

Signal Transformer
500 Bayview Avenue
Inwood, New York 11696

Schott Optical Glass Company
400 York Avenue
Duryea, Pennsylvania 18642

Teledyne Philbrick
Allied Drive at Route 128
Dedham, Massachusetts 02026

Thorn EMI Electron Tubes, Inc.
23 Madison Road
Fairfield, New Jersey 07006

Willmann-Bell, Inc.
P.O. Box 35025
Richmond, Virginia 23235

# INDEX